The Secret Life of Cities

The Social Reproduction of Everyday Life

Pearson Education

We work with leading authors to develop the
strongest educational materials in Geography,
bringing cutting-edge thinking and best
learning practice to a global market.

Under a range of well-known imprints, including
Prentice Hall, we craft high quality print and
electronic publications which help readers to understand
and apply their content, whether studying or at work.

To find out more about the complete range of our
publishing, please visit us on the World Wide Web at:
www.pearsoneduc.com

The Secret Life of Cities

The Social Reproduction of Everyday Life

Helen Jarvis
University of Newcastle upon Tyne

Andy C. Pratt
London School of Economics

Peter Cheng-Chong Wu
Chung Hsing University, Taiwan

An imprint of **Pearson Education**

Harlow, England · London · New York · Reading, Massachusetts · San Francisco · Toronto · Don Mills, Ontario · Sydney
Tokyo · Singapore · Hong Kong · Seoul · Taipei · Cape Town · Madrid · Mexico City · Amsterdam · Munich · Paris · Milan

Pearson Education Limited
Edinburgh Gate
Harlow
Essex CM20 2JE
England

and Associated Companies throughout the world

Visit us on the World Wide Web at:
www.pearsoneduc.com

First published 2001

© Pearson Education Limited 2001

ISBN 0 130 87318 7

British Library Cataloguing-in-Publication Data
A catalogue record for this book is available from the British Library

Library of Congress Cataloguing-in-Publication Data
Jarvis, Helen.
    The secret life of cities: the social reproduction of everyday life/Helen Jarvis, Andy Pratt, Peter Cheng-Chong Wu.
        p.   cm.
    Includes bibliographical references and index.
        ISBN 0–13–087318–7 (pbk.)
        1. Cities and towns.   2. City and town life.   I. Pratt, Andy C.   II. Wu, Peter Cheng-Chong.   III. Title.

HT151.J27   2001
307.76—dc21   00–053022

Transferred to digital print on demand, 2008

Typeset in 11/12pt Adobe Garamond by 35
Printed and bound in Great Britain by CPI Antony Rowe, Eastbourne

Contents

List of photographs

Photographs in chapter headings

Photographs in text

List of figures

List of tables

Authors' Acknowledgements

We would like to extend our appreciation to all those who have provided technical assistance in the production process: Matthew Smith, Louise Lakey and Stephen Pepper. Also thanks to Ann Rooke of the University of Newcastle Geography Department for her sterling efforts in processing maps, graphics and photographs to a tight deadline. These appear in a number of chapters.

There are a number of people Helen wishes to acknowledge who have contributed time and ideas to projects included here. Thanks first to Andy for supervising the PhD thesis research, which forms the basis of Chapters 4 and 5. Simon Duncan nurtured the earliest idea of the household lens used in this and subsequent research. Thanks also to Mark Kleinman and Diane Perrons for insightful comments at key stages. Colleagues too numerous to mention have contributed to a stimulating environment in which to do research. Members of the Women and Geography Study Group (WGSG) in particular have provided valuable support over the years. A particular debt of gratitude is owed to the household members interviewed in London, Manchester and Portland, who generously permitted access to their homes and their personal accounts of family life. On a more personal note, Denis Jarvis provided invaluable assistance preparing and relaying US data, contributing towards research reported on in Chapter 7. Family and friends on both sides of the Atlantic are acknowledged for their role in providing lively conduits of local knowledge and support. Helen wishes to acknowledge funding provided by ESRC award no. R00429434382 (for 1996 research) and award no. R000271085 (for ongoing UK–US research) as well as that provided by HSBC Holdings, awarded through the RGS-IBG Small Grants competition.

Andy would like to thank James Simmie and his (then) colleagues at the Bartlett School of Planning, UCL, for proposing a book on London, which initially stimulated his thinking on this topic. In particular, Stephen Merrett and David Banister with whom he tried to fashion an integrated account of 'industry, transport and housing in London'. Out of this collective work came an initial version of the employment–transport–housing relationship. The project never came to fruition and eventually appeared as separate chapters in the book that James edited for us (see Simmie 1994). Thanks to Helen, whose ideas, especially on the notion of the 'household lens', have been so productive in this project. Thanks also to Peter, whose concern for relating social reproduction and the environment has been a challenging objective for us all.

Thanks also to Ros Gill for numerous constructive comments, and Thomas and Katarina, who were a constant reminder of the need to balance home and work, social reproduction and production.

This book was written at a rather difficult time in Peter's life. After receiving his PhD degree from LSE and starting a teaching job in Taiwan, everything seemed so bright. However, at that very moment, Peter's mother was diagnosed as suffering from lung cancer; she passed away eight months later. Peter would like to say how the completion of this book reflects the help he has received from his friends and family. Peter thanks Andy and Helen for the support that they provided in the preparation of the manuscript so that he could spend more time with his mother, as well as holding down his teaching job. Peter also wants to extend his thanks to the colleagues in the Department of Landscape Architecture at Chung-Hua University for their friendship and encouragement. Finally, Peter would like to dedicate this book to his mother and brother, Kuo-Yi, who gave up his job and took full-time care of their mother in the last few months of her life.

Helen Jarvis, Newcastle, UK; Berkeley and Portland, USA
Andy C. Pratt, London
Peter Cheng-Chong Wu, Taipei, Taiwan/ROC

Publisher's Acknowledgements

We are grateful to the following for permission to reproduce copyright material:

Figures 3.1a and 3.1b from *Metropolis Now: London and its Region*, Cambridge University Press, (Hall, J.M. 1990); Figure 3.2 from High accessibility and town centres in London, 54, in *URBED (Urban and Economic Development Group)*, Greater London Authority, (the former London Planning Advisory Committee), 1994.

Photograph 1.3 Yellow School Bus (PA Photos); Photograph 3.2 'Metroland' from The London Transport Museum; Photograph 7.1 Organize Tech Temp Workers from *San Francisco Chronicle*, (Photographer: Steve Ringman).

Haringey Council Road Safety Training for use of the 'walk to school parents pledge' and associated information.

Whilst every effort has been made to trace the owners of copyright material, in a few cases this has proved impossible and we take this opportunity to offer our apologies to any copyright holders whose rights we may have unwittingly infringed.

Chapter 1

Only connect: everyday life in the city

Hang on, I'm breaking up

As we write this, the first mayoral election for London is in full swing. Beyond the usual display of political party infighting and point scoring one issue has been a permanent feature: transport. Scratch any urban dweller and not far below the surface it is possible to find a concern about transport. We have moved into an era where we are not simply concerned with a trip to work and back but with the multiple journeys that have become not just desirable but necessary in order to sustain our lifestyles each and every day. It is not an exaggeration to suggest that much of our life in cities is bound up with issues of how to get somewhere. However, getting somewhere is no good unless it is at the right time to meet the right person. Consequently, we need to consider both connection – timing and coordination – and movement if we are to capture the everyday concerns of getting by in the city.

Find a place on any train leaving the city after work and you are likely to be assaulted by a cacophony of mobile phone users making calls. A simple analysis of the content of these, and many other mobile phone conversations, reveals what at first hearing is a crushing banality: 'I'm just leaving the station now'; 'I'll be with you in forty minutes'. Mobile phones seem to have been taken up and used by many people as an active means of time–space co-ordination. Diaries have, of course, been in use for decades to fulfil such a function. What is different here is the possibility, or the need, to actively synchronise on the move. The suggestion, raised by so many daily practices and actions, is that to get by in the city requires an enormous amount of effort; part of that effort is in actual movement and another part is in the management and organisation of our personal time–space.

It is not surprising to learn that people worry about the failure to connect with others, or that they experience the constant stress of making connections. As most people would acknowledge, some stress is beneficial; however, the problem of city life is how to make the most of what the city has to offer in terms of experience and opportunity, yet not to be driven to illness in the process. Stress is often seen as a yuppie problem; however, as research has pointed out, stress is most likely to be experienced as detrimental when the subject is not in control. Thus, stress is particularly dangerous for, and prevalent amongst, lower-paid workers who characteristically have little autonomy over their work schedules. In the context of the issues raised here, the best example of the stress and lack of control that is part of urban life is being caught on a stalled underground train in a tunnel between stations. The casual observer can experience the tension and stress as minutes tick away and it becomes clear that appointments will surely be missed. What is more, the contemporary solution, to make a phone call, is not available: mobile phones do not work underground.

The main themes

This book is about the relationship between home and work in cities and regions; what is called the 'home–work balance' in the USA. Put simply, we are concerned with the fact that many people do not live close to where they work. People do have some choice over their home location, but they are constrained, to a greater or lesser extent due to income, and by a range of other factors such as housing costs and availability, travel costs and commuting time, as well as the similar considerations of others that they may share their household with.

A small degree of mismatch between home and work is normal and most people are prepared to engage in a complex trade-off between the various social and economic costs and benefits of a variety of locations. For example, a person may compromise on a small flat so that they can live in the centre of a city, to be close to work and a particular social life. Another may choose to live in the suburbs where they may trade off a longer journey to work for a house, and maybe a garden, and a different lifestyle. However, in many large cities around the world there is a worrying trend that people are no longer able to be close to their current workplace, or even a potential future one, either because they cannot afford the housing costs or because of absolute lack of availability. The daily commute has been stretched for an increasing number from the suburb to neighbouring, or even remote, cities. A return journey of 200 miles, or 4 hours of travel, is no longer exceptional.

Close is a relative value, but the point is that commuting becomes a burden and an obstacle at some point – albeit different for various individuals – as distance and time increase. Beyond such a point the consequences are significant; for example, social and health costs that affect commuters, their friends, dependants and co-workers, as well as the energy use and pollution generated in this extra travel. On top of these are the problems of those who are unable to access work because there is none within reasonable travel mode or time. Consequentially, both unemployment and skill shortage can result from this problem.

The innovation of our book is not so much identification of this issue as it is the particular focus that we develop, drawing attention to the social and organisational dimensions rather than the environmental and technical constraints. So, we would place our argument in critical opposition to those who concern themselves with environmental sustainability and with technical solutions such as 'compact cities'. Our dispute is not with the 'unsustainable' city; rather, we agree with those who point to the social underpinning of environmental sustainability. Environmentally sustainable practices can only be developed with the grain of social practices. If we are in socially divided and dysfunctional relationships, we are unlikely to be able to develop or adopt radical reorientation of our lives to harmonise with environmental objectives.

We argue that our lived relationship with home, work and recreation is at best stressed and at worst dysfunctional and damaging. In such a context we have little choice but to pollute the environment and engage in what might in an ideal world be considered as wasteful commuting and travel. The current debates that recognise the result, but fail to apprehend the process, target their attention on curbing our actions by, for example, increasing the tax on fuel, introducing toll roads or planning smaller, compact cities with housing close to workplaces. Our response is that people have to travel to get to work or see friends; while home and work zones may be proximate, if a person's job is somewhere else, or constantly shifting, they have to travel.

Although so much is relatively simple to point out, what is less straight forward is to understand exactly why it is so difficult to reduce the need for long-distance travel to work or socialise on a regular basis. While a lot of academic analysis and policy data collection is carried out on the issue of movement, it tends to have two flaws. First, it is primarily concerned with origins and destinations. Second, it is focused on the individual. The classic manner of capturing this is through the idea of the 'travel to work area'. This is the hinterland of a city from which commuters travel on a daily basis. This is undoubtedly a useful tool, but a limited one for exploring the problem at hand.

Our approach is to begin with the diversity of individual travel experiences and the surprisingly creative manner in which people routinely resolve the problems of home and work and the rest of their lives. In so doing, we begin to highlight why it is so difficult to change this behaviour. There are two points here: first, we tend to multi-task on our journeys – we may stop off at a shop on the way from work; second, our journeys, as well as our choice of work and home, have to be integrated into the equally complex patterns of movement of those we share households with. For example, people may make job choices on the basis of a desire to live in the same city with other household members, and to be in a particular school catchment area, or to be near friends. Obviously, all of these factors affect, or configure, movement. As may be expected, detailed information of this sort is not routinely collected. We have had to undertake unique data collection in order to write our book on what we term the 'secret life' of cities. 'Secret' is an ironic term in a sense: it is not a secret to every one of us in our everyday lives; however, it is a secret in terms of data collection and analysis, and in terms of the policy responses to it. Our aim here is to make the secret public, and hopefully to encourage others to think in a similar way about cities.

A related issue that our argument contributes to concerns the current interest in integrated transport systems and 'joined-up thinking/policy-making'. Integrated transport is a logical, and rational, goal of transport provision. People should be able to physically interchange between systems so that a number of modes can be used to satisfy their travel needs. This is no small feat when transport systems have developed at different times and in many cases been run in competition to one another. Yet bus and rail interchanges *are* becoming more common.

Remarkably, it is only relatively recently that the social/regulatory elements of integrated transport have been elaborated. This is not just an issue of coordinated timetabling, but also 'cross ticketing', which allows the traveller to have one ticket across the whole system. The 'travel card', which allows cross ticketing and repeat travel for a time period (usually a day), was only introduced by the Greater London Council in the early 1980s. It is still an innovation that has yet to penetrate all major world cities. Admirable though these integrated plans are, they only work with a very narrow notion of integration. The view developed in this book argues for the need to open up these debates considerably.

A recent concern of the Labour administration in the UK has been with 'joined-up government', by which it is meant that separate departments and ministries should work together to deliver policy, which cuts across traditional boundaries. This is a welcome aim, and initiatives such as the Social Exclusion Unit have demonstrated how this notion can be developed. However, joined-up thinking has not impinged upon the urban problem beyond the Urban Task Force (UTF) report (1999). Here, the solution has been cast as the compact city. As critics have pointed out, the UTF is concerned with new home building: it does not address two crucial issues. First, it is not concerned with where work will be; second, social reproduction, and in particular the whole issue of quality and access to schooling, is excluded from the analysis. We touch on these debates in relation to the UTF in Chapter 2. Our point here is to stress our radical departure from the current aspirations of policy-makers based upon our concept of the secret life of cities.

Our main conclusion, developed in the subsequent chapters, is that we need a whole new way of looking at the problems of movement and social and environmental sustainability in cities. Our point is to start with the individual in their social context. We need to understand the possibilities and constraints set upon that person by their relations to others; this we term a *situated understanding*. To this we add a renewed appreciation of the physicality of moving about cities and the absolute constraints of the body only being on one place at a time; this we term *embodiment*. Individuals operate within broader parameters of social organisation and physical arrangement, which we term *institutions*. Weaving these themes together we produce a nuanced perspective on getting by in the city.

Mismatches and their consequences

In this section we want to begin by illustrating the nature of the problem, as we see it, in more detail. Our shorthand term for this is the notion of the 'mismatch'. Mismatch may be of supply and demand, or of desire. Mismatches also imply conflict. We have two general points about mismatches. First, that they tend to be resolved in an ad hoc manner. We see this as the 'sticking plaster' solution, that is, a stop-gap, not a lasting solution. Second, that solutions are increasingly being pushed out of the public realm into the private. In a scenario where employment opportunities are relocated further from where people live, the sticking plaster is to travel further. It is a solution, but the side effects are pollution, congestion and wasted time; issues that impinge upon the wider public as well as the workers and their households. The state may encourage public transport operators to provide mass transit to facilitate commuting; or, more likely today, people will be forced to use private cars. However, this is only the tip of the iceberg.

We are witnessing an ever increasingly complex web of interactions between home and work and, commonly, more people commuting longer distances to work. The dislocation of home and work has many supporters; the notion of the suburb conjures up such an idea. On the other side of the argument we can point to the social imbalance of such communities and the relative disadvantage of many dislocated residents (traditionally women and the elderly). Politically and economically such separation can be problematic due to the fact that people pay taxes where they live. As commuting distances increase, so does the likelihood of the commuter working in a different administrative district to that in which they live. This has led many urban cores to being either close to or declared bankrupt.

The social and economic pros and cons of cities are well rehearsed. Many authors point to the empirical fact that urbanisation is becoming the majority experience for the world's population, that cities are getting bigger; bigger than they have ever been before. Urban analysts have pointed out that the processes of urbanisation are changing, particularly as cities become more integrated into the international economy and society; relationships between cities in different parts of the world have become particularly important, leading to competition between major cities.

Perhaps the most obvious consequence of this state of affairs is the impact on the environment. Urban areas have commonly spread, or sprawled, over huge areas of land, first consuming the market gardens, then the prime agricultural land that formerly fed the city. Today, food has to be transported further, wasting more energy and generating more pollution. Increased commuting implies more travel. Some of this takes place by mass transit systems but, commonly, much of the excess is solved through the use of the private car. Primarily cars (but also buses, lorries and trains) use a large amount of energy and produce a significant amount of pollution. In fact one of the major culprits of global warming is the internal combustion engine. Additionally, the sheer numbers of users of the transport systems lead to overcrowding and congestion, and wasted time.

The 'school run'

The nature of school-place allocation and quality of provision plays a partly hidden, but significant, role in the home–work problem. A useful illustration of some of the issues is provided in an article responding to John Prescott, transport minister, and his apparently laudable plea for parents not to use the car to take children to school. (See also the 'walk to school parent's pledge' below.) Cited in his support were pollution and safety cases (targeted at children), as well as the fact that there has been a huge increase in children being ferried to school by car whereby an estimated 80 per cent of the morning rush hour traffic was 'school run' related. Moorhead (1999) responded with two points. First, that legislative change in the early 1990s created the possibility of children to attend 'non-local schools' instead of the local catchment school. The evaluation of schools by 'league tables' created additional pressures for the 'school run', as well as moving house (although with rising prices this was prohibitive for many). Second, growth in female participation in the workforce has caused many women (and men) to drop the children off 'on the way' to work. Moorhead, like most parents, recognises the fact that the 'school run' is not ideal, environmentally or socially, but feels trapped by the conflicting demands of work, children's education and time (see Photograph 1.1).

Walk to school parent's pledge

The Haringey Road Safety Training Group is actively promoting this event in partnership with Haringey Education Department. As a parent/guardian you have

Photograph 1.1 The traditional 'school run': mothers escort their children from the school gates in Manchester (Source: Helen Jarvis)

an important part to play with regard to the health, fitness and road safety of your child/children. Walking to school will help to produce a cleaner and healthier environment. I fully appreciate that some parents/guardians drop their children at school as part of a further journey and, in this case, I would request that you park your car 100 metres away and walk the rest of the way to school.

1 I commit to walk to and from school with my child/children for at least the duration of Walk to School Week.
2 I commit to walk my child/children to and from school at least once during Walk to School Week.
3 I agree to park my car at least 100 metres away from the school entrance to reduce congestion and concentrated CO_2 pollution outside the school for at least the duration of Walk to School Week.
4 I commit to discuss with my child/children the benefits of walking and safe crossing behaviour.
5 I agree to help my child/children to find out about the effects of exhaust emissions and general car usage for short journeys.
　　　(Adapted courtesy of the Haringey Council Road Safety Training Group)

Homes for the workers?

House price inflation has been a notable aspect of the UK in the last 25 years, especially since legislative change to encourage near universal home ownership was enacted in the 1980s. The decline in the availability of rented property has pushed many into house purchase. One of the consequences of this has been that fluctuations in house prices can lead to 'negative equity traps' if market prices fall below past purchase prices. Clearly, this can be a significant problem for a worker losing a job: the option of moving to a new area is not possible, so travel is required to get a job to pay the mortgage for a house that is in the wrong location, but the worker can't afford to sell.

Urban areas, especially London, have experienced particular problems due to the dual pressures of employment growth and a limited supply of new housing. One of the consequences that has recently come to light has been the relationship between wages and the size of mortgages required to purchase a house. Brindle (2000) notes that London entry-level house prices are £97,000. The average calculation is that loans are allowed up to three times annual earnings. A typical public-sector worker, such as a nurse, will earn only sufficient to be able to afford a £53,000 mortgage. The number of rental units has fallen by 13 per cent since 1983, and rental costs are also very high. Rental can often be more, per week, than an equivalent mortgage repayment. For example, average rentals are around £275 a week, and a repayment on a £97,000 loan is approximately £200 a week. The catch is that you have to have a larger salary to qualify for a mortgage. Nearly 200,000 are on waiting lists for subsidised social housing. Travel costs are high too, so that the option of living in the suburbs and only slightly lower housing costs has to be traded against travel costs, and time and inconvenience in commuting. A weekly travel card for the inner suburbs is at least £25 a week. The consequence is that public-sector workers and support staff in the private sector, generally, cannot afford to live in London. This in turn is generating a labour supply problem.

A recent response by government, one copied from places such as Aspen in the USA, has been to suggest 'interest-free top-up loans' for 'key workers' in the public sector, as well as looking at cheaper housing supply (Hetherington 2000). Hetherington notes that while the problem is most acute in London, it is replicated in many of the metropolitan areas of the UK. Caught in-between are the middle classes who aspire to larger houses or better schooling and environment and seek out cheaper property not just in the suburbs but in urban areas between one and two hours' travel away (see Langton 2000). Similar properties in a smaller town (for example, Milton Keynes has a 40-minute commute, with an average house price 50 per cent that of London) may be a quarter of the price of those in London; even the extra commuting cost pales into insignificance with respect to total outgoings saved by such a move. Once again, the real cost is travel, time, pollution and stress.

'Off the edge' cities

The consequences of this process can be seen by reference to developments in the USA. Ed Soja (2000: 260) gives a graphic example of the extension of this process with what he terms the development of 'off the edge' cities. His reference here is to 'edge cities', which are akin to small nuclei focused on retail malls and business parks (see Garreau 1991); 'off the edge' cities are anticipatory cities/suburbs beyond the edge of the city proper and based upon new job growth. As job growth has stalled, the 'jobs-housing mix has become disastrously out of whack'. Soja discusses the fate of Moreno Valley, 60 miles east of downtown Los Angeles: '[m]any working residents are forced to rise well before dawn to drive or to be taken by a fleet of vans and buses, often for more than two hours, to the places of employment they held before moving to their affordable housing.'

Hypermobility and environment

There is a great irony in the fact that as we enter the age of the Internet, where it is claimed that 'geography no longer matters', our travel and mobility patterns are increasing in volume and complexity. Physical presence still matters to most people, either in their employment lives or their non-work lives. We desire to meet others in person. However, in our cities, the costs of doing this – personal, environmental and economic – are increasing. Our great transport systems were developed to link people's home and work and, commonly, it was envisaged that this would involve a radial commute, twice a day, followed by smaller trips around the home node or the work node. As we have noted, this pattern of home–work relationship has changed: on one hand, trips have got longer; on the other hand, they have become more complex centripetal journeys or multiple trips. This has led to a massive modal shift to cars, and to a rapid exceeding of road capacity. It has also led to a huge growth in mobility: we travel to work, to the shops, to see friends. Invariably, the journey that may once have been a short walk now involves either a car, bus or train. John Adams (1999), in a report for the Organization for Economic Cooperation and Development (OECD), sums up this process as one of 'hypermobility'.

Adams also outlines some of the consequences beyond the obvious environmental degradation and loss of land to roads. He points to the significant evidence on the link between transport-generated pollution and the incidence of asthma, especially in children. He also highlights how social patterns of play and interaction are disrupted by the dominance of car travel. Children, he points out, are becoming more unfit, and more overweight, due to the lack of exercise in walking to school and the lack of outside play. The decrease in the latter has been attributed, in part, to the danger of being near traffic. More generally, communities are isolated by their dissection by difficult-to-cross roads. Finally, local services, such as shops, relocate to facilitate bulk-buying by car-users. While some commentators have suggested that supermarkets and shopping malls are the new community spaces where we meet and interact, this does neglect the fact that not everybody has access to these places, or that for some there is a considerable cost to their use.

This picture will not be new to readers; we have heard these problems much debated in recent years. Alongside them has been the question of what can be done to reduce energy usage and pollution emissions. Here, we argue, the scope of enquiry has been too restricted or inappropriate. Proposals abound for encouraging a shift between transport modes, from cars to buses or trains. However, the mass transit option is not always available because of such things as capacity constraints, costs or network limitations (it doesn't connect A and B). On the other side of the coin a sophisticated method of road pricing and parking tax is also being explored.

We do not want to enter into the debates about financing and taxation; we want to suggest that these responses may develop from an inappropriate assumption as to why and how people travel. The pricing debate always gets stuck on the question of regressive taxation that hits the poorest section of the community. The question that we believe needs to be asked is profound: why do people travel at the times, by the means, and in the manner that they do? If we could understand this question then we could move on to issues such as: how can we reduce that need to travel?

Our point is that this question is not simply one of home–work relationships; it is about the interaction of housing and labour markets and the configuration of transport systems. It is about the nature of schooling access and provision, as well as the form and location of retail and leisure facilities. Moreover, it is about the form and nature of personal support networks, from household members to babysitting circles, from local car and house repairers to friends and relations. It is this whole complex web, within which travel and mobility is located, that we want to emphasise. Thus, we argue that it is necessary to view mobility as a *situated* event. Change may be achieved and blocked in a variety of ways; ways that are not reflected in the crude attempts at pricing solutions, or relocation, that are currently on offer.

Characterising the problem: four dimensions

Thus far we have sought to provide an illustration of the consequences of ignoring the secret life of cities in the way that we run our cities. We will argue, later in the book, that we have much to learn from the ways in which ordinary people 'get by' in the city. In this section we begin to sketch out our conceptual foundations, our way of getting to know the city. We do this by dividing it into four dimensions: employment, housing provision, movement and social reproduction. We discuss these topics in detail in Chapter 2.

The home–work relationship is necessarily concerned with transport and movement issues, as well as the availability of jobs and houses. Transport, housing and employment are all subject to complex allocation and provision processes summarised by the shorthand 'market mechanism'. As we know from any standard economic text book, markets resolve supply and demand for commodities. Markets have imperfections, which are commonly attributed to imperfect knowledge, but there are further barriers to 'market clearing'.

The classic case here is of sluggishness in market transactions. The example often used relates to geography: a vacancy exists in New York, and an unemployed person lives in Chicago. The person could move, but it will involve considerable social and economic costs and thus might not happen. In labour markets these are referred to as 'frictional' factors, and they offer good reason why applied economists concern themselves with 'local labour markets'. Economists would also admit to 'structural' factors. In the case of labour markets this is characterised by skill and occupational segregation: a vacancy exists for a merchant banker, but the unemployed person is trained to be a chef. Thus, it is possible to have simultaneously both unemployment and job vacancies. The boundary between structural and frictional factors evaporates the more closely the problem is inspected: retraining could conceivably, in time and with investment, relocate a person from one segment of the labour market to another; likewise, relocation from one place to another is also possible.

The point here is that while in (neo-classical) theory markets operate, in practice there are problems. These problems mean that 'labour markets' operate in a more structured fashion; they can be considered to be segmented by location and occupation. The most important social factors (beyond those already mentioned) associated with the limitation on 'free movement' in labour markets are distance (or more accurately travel time), home location and household living arrangements. These add up to a fearsome barrier to free movement: some barriers are self-imposed, and some are imposed from elsewhere. Cutting across housing, labour and transport allocation and provision are the dynamics of the social unit, the household, and its configuration in gender, class, race and generation. If we add to these 'factors' the point that housing markets, and transport provision, also have their own internal dynamics, the picture becomes even more complex.

It is this complexity that we are seeking to address in this book. It is clear that they are neither fully described by the notion of the market nor likely to be resolved through the price mechanism alone. As a preliminary step, the remainder of this section briefly introduces the characteristics of what we consider the 'four dimensions' of city life.

Employment

In the past, urban and regional analysts focused their attention on understanding the structural conditions that generate home and work locations. There is a long tradition of town planning, mainly through land-use zoning, that has sought to offer rational solutions to the competing demands of sanitation, nuisance, and social and economic well-being. Underpinning this practice is the assumption that home and work zones can be close, and easily linked via efficient transport connections. Supporting such arguments are analyses that focus on labour markets and which note how firms are often content to grow *in situ*, rather than move. Thus, as they grow, they draw in more employees. When this is replicated many times over, the local labour market is exhausted and people are attracted from further afield. These new workers may choose to move or simply commute. If the latter, the city region expands: its travel to work area gets bigger.

The patterns of growth and change are not always smooth. Analyses of the post-1970s era have pointed to the process of deindustrialisation, which have resulted in manufacturing industry relocating from core city regions, and even from core economies, to peripheral locations. This led to considerable *in situ* job loss. New economic activity in the form of the service sector arrived, creating many new jobs. However, those that lost their jobs, commonly men, seldom obtained the new jobs, which were secured by new entrants into the labour market, and often these were women. Those that lost jobs became unemployed, retrained or found similar jobs in new locations. This latter group either commuted or moved house.

Housing provision

The home side of the equation has not been smooth either. Housing has the characteristics of being a 'lumpy' development: it is created in big, costly chunks and exists for a long time. Thus, it brings with it a fixity of location that can quickly become out of sync with housing needs. Housing also has an entry barrier of either social regulation and/or cost. In a housing market dominated by rented housing, mobility is limited only by the social desire to move (see below) and price. Price is linked not only to house size, but also location.

Again, with location, it is not a simple distance decay function but one that is mitigated by style and neighbourhood characteristics, an important aspect of which is the current social composition. A location out of town in a fashionable area may command a higher price than an unfashionable one closer in. Additionally, location is best thought of not only in terms of pure distance, but also in terms of time. One may be close to public transport but, if the network is inefficient, it may be better to be further away in real terms, where the efficiency of the transport link is more favourable.

The rental housing market has traditionally been subdivided into social (or subsidised) housing and free market housing. Social housing has, historically, been built in specific locations, often to resolve home–work problems: to provide affordable housing in the right location. However, due to the economics of social housing provision it has commonly been built on cheaper land, normally that which is either less accessible or less desirable to the general population.

The shift away from rented housing to owner-occupation was a developing trend in the latter part of the twentieth century in the UK and elsewhere in Europe. In the UK, the housing market experienced a seismic shift in the 1980s with the so called 'right to buy' legislation. This enabled people living in social housing rented from local authorities to buy that property, which rapidly led to a rise in the proportion of the total housing stock in the owner-occupied sector. A free market in housing for owner-occupation does at first sight simply regulate access in terms of price; however, in practice this is complicated by access to credit (a mortgage). In addition, there are market fluctuations in property values.

Photograph 1.2 The 'living wage' debate has taken hold on both sides of the Atlantic (Source: Helen Jarvis)

In the 1990s, in the UK, this gave rise to the phenomena of 'negative equity' for many new house owners. Put simply, the market price for a property fell below that for which the mortgagee holds a mortgage. So, if a person wished to sell their property, they would not be able to clear their mortgage debt; in short, they were selling for less than they bought. Given the high level of house prices in relation to income levels, it was beyond the means of most to simply pay off such a sum of negative equity. For example, on a purchase of £50,000 a drop of 10 per cent would mean a debt of £5,000; such a sum could easily account for between one-third and one-half of the annual take-home pay of a household. Thus, negative equity had the power to lock in people to particular houses. Due to high loan rates, constant employment was also necessary; thus, inevitably, as people changed jobs they had to travel further. The alternative was unemployment, and loss of the house via repossession.

Putting aside the problems of negative equity, the problem of limited rented accommodation and house price inflation relative to income has generated another set of issues. Basically, it is not possible for people on average wages to afford a mortgage in London (see Photograph 1.2). Of course, many bought property at lower prices and have lower mortgages; new entrants are not so lucky. Such a process has a huge impact on labour markets, generating shortages in many jobs. As we noted earlier, a recent UK policy has established a special top-up loan for key public-sector workers (COI, 2000).

In a more detailed analysis of London, Pratt (1994a) pointed out the complex transport relationships that housing and labour market structures generate. The fortunate can live close to work, or commute directly in. The less fortunate have to work 'against the grain' of the transport system, living in social housing in the city and commuting across or outside the city.

Movement

Transport infrastructure is even bigger and more 'lumpy' than housing. It always suffers from the problem of being out of date almost as soon as it is complete, due to the pattern and rate of change of its users. Somehow, transport infrastructure has to be modified over time; it is seldom possible to build so far in excess of capacity that overcrowding does not ensue.

While there are some new dimensions to the problem of urbanisation, the perennial problem is common – that of growth outstripping resources. This is manifest in the ways that infrastructural capacity in either utilities or building stock is exceeded. Associated with this is the problem of renewal: new growth literally overtakes old; sometimes the old is knocked down and replaced but, more commonly, due to cost, development flows around, or over, old development. The result is not only a patchwork of land uses, but also the coexistence of infrastructures tailored to different needs. David Harvey (1989) has captured this well with his notion of the city as a palimpsest, where the old uses bleed through into the current ones and intermingle in a complex pattern. It is in this complexity, rather than on the neat planner's land-use map, that people live their lives.

In London, as in many other cities, the main transport infrastructure was laid down more than a century ago. Even then, many of the new service networks merely replicated existing bus and tram routes laid down many years prior to that. Aside from the problems of repair and decay, the network is less suitable for current needs than it might be.

The problems of the network are most commonly expressed in terms of capacity constraints, exacerbated by the rush hour (which now extends for much of the day). A new phenomenon has been noted regarding road building, as illustrated by the example of the M25 orbital. The problem is that new roads, or new road space in the form of extra lanes, generate more traffic: more roads generate more congestion (Banister 1992).

However, what is less often addressed is the issue of the network topology constraint. Adding new roads, or new underground services, is extremely disruptive and costly, as Londoners have noted with the construction of the Victoria Line and, more recently, the Jubilee Line. The 'cross-rail' tunnel link that would provide an east–west corridor linking Paddington and Liverpool Street for overground trains (like the RER system in Paris) is still under discussion, as is the Hackney to Chelsea north-east to south-west tube link. However, such proposals only scratch the surface.

The guiding assumption behind original tube lines was of the daily radial commute; however, as jobs and housing have developed a more complex relationship, as well as travel demand more generally (see below), the demand for travel has become centripetal too. In effect this mismatch between infrastructure and network supply and daily travel demands has forced people to move to road transport, mainly cars, as the road network is better configured for such travel. This network topology, coupled with the nature of transport provision and its cost, creates what we term the 'grain' of the city. There are some corridors and places that are easier, and quicker, to travel down and to; the alternative, cutting across and against the grain, requires far more effort and time.

Social reproduction

The fourth driver of urban change is what we term social reproduction. This term covers all of those activities that make up our everyday lives yet are not covered by the home–work relationship. In a sense this is our underpinning dimension that cuts across the others; it is the aspect most ignored but the one which, we feel, has most to offer in terms of real 'joined-up thinking' about this problem. We draw inspiration from feminist critiques of economic analysis, and the debates about the value of reproductive labour. We take from these debates a clear signal that reproduction is of equal importance to production: they are codependent. Thus, to look at the city in terms of work alone yields only a partial view. The worlds of non-work (or unpaid work), shopping, educating and recreating, along with their associated forms of transport, must all be drawn within our analysis. We would see production concerns (traditional home–work relations) and reproductive concerns as co-constitutive; both are embedded within one another (see, for example, Wilkinson 2000).

The social significance and demand for reproductive activities has grown enormously, both with the increase in leisure time in the second half of the twentieth century and with the growth in disposable income. The timing of our consumption has also changed, both in the life-course and within the day. With retraining and lifelong learning, education is no longer complete in the early teenage years but commonly continues well into adulthood. The development of longer shopping hours has led to the emergence of 24-hour trading. While this currently affects a minority, there has been a huge stretching of the 'shopping day' for the majority of people; this also applies to other recreational activities.

The location of these activities has increasingly shifted from local neighbourhood facilities to 'out-of-town' centres for retail and leisure. In respect to compulsory-age schooling, other dramatic changes have occurred, not simply in the growth of private provision, but also the regulatory change that allows children to attend a school that is not closest to them. In inner-city areas, where education provision is generally very poor, extensive and complex patterns of individual commuting take place so that children can reach the school of their parents' choice. The bussing of children that takes place in the USA to create 'balanced' schools seems insignificant when compared to such movement. Such is the traffic generated from this activity that it has become a transport policy issue of major concern (see Photograph 1.3).

Photograph 1.3   The US-style 'school run' imported to the UK (Source: PA Photos)

Finally, the composition of those participating has changed too. The traditional identification of sex with social role has broken down, or at least been modified. More women now work outside the home and in some cases women have become the 'breadwinners', but more commonly we now have more multiple-person working households. The demands of the '24-hour society' have meant that, increasingly, household members can be working on different shift patterns (see Demos 1995; Christie 2000).

We argue that the household, as a key site of mediation of social reproduction concerns and responsibilities, should become a new focus for analysts of the urban problem. If one can get inside the household and understand why they live where they do, and why they travel, one might get a step closer to understanding the current configuration of the urban home–work problem.

The road to the present: the dynamics of urban growth

The previous section provided an initial outline of the key factors of change; perhaps they could be envisaged as the four motors of change. In the opening to the chapter we signalled that a concern with time–space also informed our analysis; it is not surprising, then, that we now seek to contextualise these factors in time–space. The point we want to convey is that the mismatches and resolutions of today build upon those of yesterday. Cities have delivered a considerable legacy to us in the form of the built environment and the broad city plan. In cities such as London the basic layout of key streets and thoroughfares can be dated back more than two millennia; it is no wonder that there are problems to movement today! Of course we seldom, if ever, have the chance to begin a city anew and in any case, if we did, we would be faced with similar issues within a generation. Thus, the most productive approach, but also the most difficult, explores solutions broadly within the context of the current physical forms.

The commonly held response by policy-makers to the local problems of inner-city economic decline, poor infrastructure or social exclusion has been a twofold one: structural and spatial. Structural policies aimed at the whole economy, or at redistribution of resources, are like the rising tide: more resources help to 'raise all boats'. However, such is the nature of public investment that the tide invariably turns before all needs have been met. This, it is often argued, is the problem of specific locations, and it is the justification for area-based remedial policies (see Smith 1999).

In parallel, urban theorists have stressed the ways in which globalisation has caused urban growth to take on a polarised character between an internationalised financial services sector at one pole and a national or regional manufacturing and servicing sector at the other pole. Spatially, this is not manifest in a simple core–periphery pattern, but a complex mosaic. This mosaic emerges from the existing infrastructure patterns noted above, as well as patterns and processes of housing, retail and leisure provision.

The complexity of mobility and stability, structural and spatial, is the character of the contemporary urban problem. It seems clear to us that simple land-use regulation or traditional structural and spatial policies, or even the relatively new pricing policies, are far too crude a tool to address a delicate problem. It is a system based upon a form of cognitive therapy reinforced by economic shocks. A more effective and a more humane solution lies, we believe, in the sphere of understanding everyday life. We need to explore the simple-to-ask (but difficult-to-answer) questions, such as why do we need to travel, and what would need to change to allow us to travel less?

Initially, jobs and work were more or less coincident in cities. Prior to large-scale industrialisation people worked where they lived, or they worked in adjacent workshops. With mass industrialisation and the coming of the factory many workers were required at the same place. In many cases factory owners created their own housing in order to accommodate the workers, who were often migrants from rural areas. As histories and novels of the period make clear, the quality of the housing was often poor, and it was almost invariably tied accommodation: if the job was lost, so was the housing (see the Manchester of Elizabeth Gaskell's *North and South* or the Preston of Dickens's *Hard Times*). There were examples of industrial philanthropists and social reformers who sought to develop 'model communities' that sought to establish a harmonious relationship between home and work life (for instance, Titus Salt, Joseph Rowntree, Robert Owen and Ebenezer Howard). In the absence of formal land-use planning, these private and community developers, and factory owners, established the foundations of many contemporary urban settlements.

This legacy is important in that, in the development of cities, a complex and fragmented pattern of private ownership of land developed. If this fact is added to that of the legacy, and amortisation, of the built environment, one can quickly appreciate the slow redevelopment pattern. Furthermore, given that the whole is interwoven by a street pattern that was itself a relic of field or heraditament boundaries then it is clear that cities are always struggling with their past. We can see some examples of the enormous and traumatic upheavals of urban life and the urban fabric that were associated with the arrival of the railways to many urban areas. For example, Porter (1994: 217) documents and graphically illustrates the scenes of physical upheaval and social disruption associated with the building of the London–Birmingham railway in north London in the mid nineteenth century.

Cities have grown in size as their social and economic importance has increased. More people are drawn into employment to work in the factories and offices. The process is one of land-use intensification. The classic example of this is the twentieth-century push of buildings upwards. More people can work on the same site. At a very basic level this increases the draw of the city on its population. It is not possible for everyone to live close to their work, even if they live in high-rise buildings or tenements. The solution is to travel and, in turn, this puts pressure on transport systems. Initially, cities were limited in size by a simple rule: the length of time that it took to walk to work. Of course, the development of new transport technologies has served to increase this, first with the tram and then the bus and train.

Whereas walking does not require more than a dirt track, powered (usually mass) transit requires specific infrastructure (roads, rails or tram lines) and an agreed route. These routes distort the city; the lines that they trace become corridors, or fingers, that extend out of the existing urban area. Initially, these corridors funnel people into work in the city. Additionally, more people can travel from more remote locations to the city. Thus, the city develops a greater hinterland of dormitory suburbs, or dormitory towns and cities. The development of London's underground lines is a good example of this. In London, as in the USA where there was also extensive rail development, the rail link, or specifically the location of a station, created a rise in land values and land speculation so that settlements grew up, like a string of pearls, along the route. Housing developers and tube operators jointly exploited their linkages. The classic example is of the entrepreneur Charles Tyson Yerkes (see Porter 1994: 317). Suburbanisation was thus fuelled. A more extreme version of this practice can be observed in Tokyo. Here, the main railway lines are developed and operated by the same company that builds the suburban houses *and* develops the huge department store at the city end of the line (many of the lines are known by the department store name).

The cost of transport was and still is an important factor. While transport operators sought to keep prices high, the government, under pressure from workers, sought to achieve concessions, such as the famous 'working man's penny fare'.

While much of the early suburbanisation extended the city, the development of new transport corridors and mass production techniques meant that factories required both more space and access to wider regional and national markets. Moreover, electrification freed the factories of the locational resource ties of water and steam. These advancements enabled development along the improved arterial road network, for example the A4 in the west and the A10 in the north-east of London. New sites developed alongside these roads, usually in commercially developed industrial zones: industrial estates. In a similar process to the tube and railway developments, these roads, and the new industrial estates, became the focus for residential development. The new factories were able to attract, and draw upon, a local labour market (see Pratt 1994b).

This is the ideal version of urban sprawl. The built-up area gradually extends, but it develops a polynucleated form rather than remaining focused entirely on the original core. Implicitly, these developments were class structured, as they were dominated by employment type. Alongside this new development often comes resistance. The arrival of the railways into north London was resisted in part by the existing landowners because they feared that it would lower the tone of the area. The fact that Kings Cross, Euston, Marylebone and St Pancras all stop at a line equivalent with the Euston Road is due to land ownership patterns south of the road. Originally, the Bloomsbury estates were gated to create secure neighbourhoods. In contemporary cities a common debate is about 'gated cities'; this is not a new phenomenon. Another tactic was employed by those living in Hampstead: initially, permission to run a tram up Roslyn Hill to Hampstead was resisted lest it encouraged access by the working classes.

In summary, what has been documented in the snapshots above is the constant struggle to overcome the constraints of space by land-use intensification and by extension. The same tensions, though not the same resolutions, can be traced in all major cities. Of course, the particular character of industrialisation and its timing, plus the dominant transport technology used and the character of land-use regulation, all shape particular cases. Intensification requires more workers to service a given location. Due to space, cost and environmental limitations, this generates commuting from the suburbs and beyond. Suburban development can be dormitory in character or linked to extension, that is, to the development of local labour markets and decentralised employment. Interwoven within this are patterns of resistance established by landowners and residents that create enclaves of development.

Ideal solutions: new towns and compact cities

In the previous two parts of this chapter we outlined what we called the four motors of change and underlined the historical legacy and context within which they operate. We will now continue this review and take in some of the initiatives that have developed to address what is at the core of the home–work problem through a particular means – land-use zoning and planning regulation. Our aim here is to emphasise the argument that we need to go beyond physical solutions, which are concerned with matching origins and destinations, and focus on the experiences and needs of ordinary people.

The debate that we want to highlight here concerns planned solutions, primarily in the context of publicly controlled town planning. The new town experience in the UK and elsewhere does not seem to have created balanced communities from a movement point of view. In this sense they seem little different to old cities, such as London, and decidedly unplanned cities, such as Los Angeles. Clearly, there are many admirable and positive aspirations in planning; however, planning legislation, even in its most developed form such as in the UK, has been quite limited and rooted in land-use regulation. As we have argued, physical design is at best a partial tool in approaching this problem. In this part we review some of the planning legacy.

New towns

The general pattern of sprawl and eventual in-fill between transport corridors, facilitated by road transit, characterises much urban development. To a significant extent the planning system was created and established in the UK to deal with the 'ribbon development' and 'sprawl' created by road-led urbanisation. The main impact of the arrival of planning in 1947 for London was to secure the green belt, a sort of *cordon sanitaire*, which would protect London from sprawl. Alongside this there were plans to decant growth to new, self-sufficient settlements: new towns.

The new towns were inspired by the garden city movement and the writings of Ebeneezer Howard. The garden cities, beginning in 1902 with Welwyn, were an explicit attempt to bring together home and work within a balanced community; already an early response to the home–work mismatch. The problem is that governments have never been able, or seen fit, to totally regulate the location of employment. So, it has always been a matter of chance if a self-sustaining community can be developed. Historically, the garden cities and, more radically, the new towns spread around London failed to hold on to their populations. Many of the workers found work in London and commuted to the new towns, thus effectively leap-frogging the green belt (see Hall 1989).

This process of transport-enabled spread through daily commuting patterns raises many problems for the city and these can be summarised in terms of governance. As the travel to work area grows there is cross-dependency of populations from one administrative area to another. People work in one place and pay taxes in another. Likewise, training or unemployment policies of one area support another. Attempts to resolve this problem have commonly resorted to strategic government, often a higher tier of government. However, cities seem to have the capacity to outrun administrations. In London, for example, the argument has been made that the south-east is the real and active hinterland and, as such, it needs to be considered as an interactive whole for the management of training, services and economic development (see Simmie 1994).

Green belt policies, and those of new towns, were, in no small part, attempts to conserve land and reduce sprawl. The response was to apply zoning strictly and to control permission for new building (planning control). However, economic growth continued apace, and it had to go somewhere. Land-use zoning was not an effective restraint. In the 1960s efforts were made in the UK to control the location of offices. However, this was more part of a national redistributive strategy rather than one that might balance home and work. Secondly, as we will detail below, economic restructuring (and flexible working and multiple careers) has led to a complete fracturing of co-location of employment and housing. The result is a polynucleated settlement pattern with multiple flows of people to and from work.

Compact cities

After the Rio summit on climate change, the European Commission produced a discussion paper to outline possible implications of the Rio agreements for urban areas (Commission of the European Committees 1990). The idea of sustainable cities is a contradiction in terms to those who suggest that cities are net exporters of pollution (see Owens 1992). The main idea that was developed was that of the compact city. The compact city idea seeks to find practical solutions to minimise energy emissions. The central idea was to minimise journey numbers and lengths, on the assumption that this would then consume less fuel. This was bolstered by arguments about energy uses for heating and cooling, as well as building materials and designs.

Urban trip patterns were analysed and the result was a central recommendation that smaller, densely settled cities would minimise journey lengths. There are a number of problems with such an approach (see Breheny 1992a and b for a critique and summary of the key research). First, the idea of a compact city links with an already existing design aesthetic of small intimate urban spaces modelled upon mediaeval northern Italian hill towns, such as Sienna, and a planning aesthetic of what has been derisively called 'town cramming'. This association led to a willing acceptance of the idea by the planning community, who, as we noted above, are strongly opposed to 'sprawling cities' on aesthetic as well as environmental grounds.

The problem with this notion is that it is based on a form of spatial determinism; that is, the idea that particular spatial forms are inherently 'social', and others are 'anti-social'. At the extreme, this idea is clearly true, particularly when there are physical barriers to movement (a key case is where a community is bisected by a major road: see Elkin et al. 1991). However, more generally, the idea is based upon likely interaction patterns and implied sociability that may result from it. There is a long debate in sociology and geography about proximity and social relations. Indeed, one of the key arguments of commentators such as Giddens is that the social relations of 'community' have been stretched over increasing times and spaces in recent years, especially in cities. The weakness of this approach is that it takes for granted a particular lifestyle and, crucially for our argument, a particular home–work relationship. Indeed, we believe that such processes have led to social cleavages within residential communities (see Chapter 2).

A second, and related, point is that the compact city model is fundamentally based upon the minimisation of diurnal radial commuting; it is also based upon an economic structure assuming a number of small market forms, that is, the idea that people use the local shops and services. Despite cogent arguments (based upon statistical modelling of trip patterns) to propose ideal city sizes and densities, the social and economic bases of their generation are ignored (for example, see Banister 1992; Rickaby et al. 1992). The compact city rests upon an asocial notion of the household, as if every individual maximised there utility (or they are able to live close enough for this not to matter).

Finally, notions of housing and labour markets do not seem to enter into the equation. As we have pointed out, the structure of housing and employment provision is not equally distributed. Thus, the compact city, or its concept, is not based upon a credible model of human interaction. Moreover, and consequentially, the tools to control and regulate its development – land-use zoning – are unlikely to be effective.

The notion of the compact city is seductive but, even so, it does seem to fall foul of another key problem: how does it address actually existing cities rather than those that we might ideally plan? The answer to this, in part, has been explored in the context of the report by the UK Urban Task Force (1999), the first major rethink of urban policy in the UK for a decade. The Task Force's agenda was rather limited; namely, how to accommodate more house building within the south-east of the country. The demand for housing, due to the jobs–housing imbalance, had been growing. House builders were seeking permission to build upon green field sites. The Government's concern was to find a rationale for 'brown field' site development to be sustainable and to aid urban regeneration. Brown field site development promotes the reclamation of derelict land and thus reduces new land take. Given that many inner-city sites are brown field, such development might also achieve a 'compact' form too. The UTF advocated new house building within cities and on reclaimed industrial land; it was against new town developments.

Predictably, the UTF's recommendations have come under considerable criticism due to their failure to take into account the factors that generate 'inappropriate housing demand'. Breheny's (1999) report for the Town and Country Planning Association (a pressure group supporting new town and garden city developments) focuses its critique on the employment issue. As we noted above, this issue was not addressed in the Task Force's report. Breheny provocatively entitled his response to the report *The People: where will they work?* Employment has been suburbanising, so urbanising housing is not going to help. Even this is a crude generalisation; it does not explore the segmentation (structural and spatial) of housing and labour markets, nor the individual problems of negotiating them.

Another criticism is that the UTF does not address the issue of education and school provision. Again, the issue is about the quality of schooling in the city. While the UTF suggest that suburbanisation is led by schooling, they wash their hands and pass it off to the Social Exclusion Unit. It is an irony because, in the early 1960s, the starting point of modern urban policy was based upon Education Action Zones. The problem today is again more complex, as it is possible to obtain high-quality education in urban areas; the problem is that it is commonly regulated by financial barriers (private schooling). If we add to this the complex relationship between area access policy and attainment, the potential to hold some people in the city, and to generate complex trip patterns for both rich and poor, is considerable.

A long and convoluted debate has been carried out regarding social exclusion and the city and a review by Smith (1999) highlights the tensions between policies targeted at structural and spatial objects. There is an acceptance that there are spatial differences in poverty and exclusion; the question is how to remedy it. Spatially focused policies are criticised for failing to address underlying problems of inequality. The alternative is structural policies that seek to 'raise all boats'. However, the reality of redistributive or structural policies is that the impact in the most deprived areas is only felt after some time. Commonly, to continue the nautical metaphor, the funding tide is on the ebb before the islands of poverty are relieved. Thus, the combination of structural and spatial policy is justified.

The solution generally offered to the 'sink area' is positive reinforcement, also backed by a strong US 'workfare'-style employment programme that encourages, and enforces, the unemployed to travel outside their locale for work. There is a general acceptance that it is not possible to attract investment into many local labour markets, but jobs may be available 'close by'. However, returning to our core agenda, we see that people must travel further distances to work. Also, we see that those who can least afford it have to travel either furthest or along the most difficult routes.

Research by Pratt and Gill (1991) on unemployed workers in a very deprived area of north London (Tottenham) highlighted that one of the barriers to escaping from unemployment was the time, cost and difficulty of travel: in this case, the problem of travelling across London when numerous modes and networks need to be negotiated. Other research has highlighted the lack of social, time-management and organisational skills that many unemployed persons have (especially if they have never worked).

Thus, what may seem to be a negotiable travel plan can seem impossible. This adds another dimension to the worrying trend that Adams discusses above, namely, that the unemployed too must be hypermobile. As the body of work on rural transport neatly illustrates (see Hodge and Whitby 1981), the combination of physical isolation, depopulation and the associated withdrawal of public transport has led the rural poor to become car owners in order to maintain the hope of getting to work. This is, in effect, a tax on being available for work. A less reported fact is that, despite apparent proximity, the urban poor are just as isolated as the rural poor from the places that they need to access. Hence, in the midst of plenty, poverty thrives. This leads us to stress a way between the spatial and the structural; our concern is to highlight the institutions and networks within which people have to operate, and which set the limits on their horizons.

These issues feed back into a final point, a theme that runs through the book: social sustainability and environmental sustainability are intimately linked (see White and Whitney 1992). Generally, the argument has been expounded, and accepted, in a development context. This point is clearly made in the Brundtland Report (WCED 1987); that is, environmental sustainability is not possible without a socially sustainable community. However, in the developed world it is often assumed that this caveat does not apply. Our point, backed up by reports by the Social Exclusion Unit, is that cities are, in many cases, unsustainable and in the process of social disintegration or, at best, social stress. While traditional explanations of social deprivation are central (primarily, absolute poverty), our argument here is that the dislocations of the home–work and social reproductive environments prejudice full participation in society or, at least, make it more difficult.

Conclusions

Getting by is a problem for us all, although it is experienced differently depending upon our unique access to social and economic resources, and a whole set of situated and institutional constraints. The current dominant debates about the home–work relationship, and urban sustainability issues, offer a complex picture and a bewildering array of solutions. Too many of these are what might best be termed 'engineering' rather than 'social' solutions. It is as if people are like rats in a maze; but people, and rats, have intelligence. If one looks carefully, a whole range of innovative and ingenious local solutions are made to the home–work problem by people in their households and communities. The problem is that these are ignored in the engineering mind-set, often working against them. Our intuition is to learn from how people go about solving their local problems and then to use this as a basis for any more general problem solving. In the following chapter we begin to set out our way of framing and conceptualising the problem. This is followed by detailed observation of a number of local solutions and what we take to be their implications.

The aim of this chapter has been to sketch out the complex patterns of interdependencies that constitute contemporary urban development. The point of entry that we chose was the home–work relationship. Any city dweller will acknowledge that this relationship is under stress: more cars, congestion, distance and time spent commuting. As we have noted above, not only is commuting in its current form inconvenient, but it also carries with it huge social and environmental problems: stress, environmental damage and pollution, health problems and social exclusion. We have noted that society has tried to manage these problems in a rather piecemeal fashion, and even then to concentrate only on the issues of immediate economic impact.

Our contention is that because of, and despite, these conditions, people have to develop ingenious strategies for 'getting by' in the city. Some groups, usually favoured by wealth and social privilege, have less difficulty because they work with the 'grain' of the city. Most others, either through direct or indirect relation, have to work against or across the patterns and flows of city life: this is the secret life of cities – how they really get by. The argument that we will develop in this book is that from a conventional perspective these strategies are neither 'rational' nor 'efficient'. Thus, even with the best will in the world, conventional remedies are unlikely to have the desired effects. We argue for the need to examine the secret life of cities more closely and to learn how it operates.

The secret life of cities is not hidden or mysterious – it is the banal and everyday. It is dropping into the shop on the way home from work; it is dropping children off at school; it is living in the best place to suit multi-earner, and multi-activity, households. It is, in sum, how we organise our everyday lives. Lest we imagine that these are marginal concerns, recall that 80 per cent of the morning rush-hour traffic is 'school run' related. The secret life of cities is the pulse and lifeblood; the formal perspective – how cities are meant to function – is the peripheral one.

This chapter has sought to elaborate our case for the complexity of city life. We have indicated that the home–work relationship is complex and that, in addition, it is articulated with a whole range of other 'moments' of city life. We have chosen to build our argument around labour markets, housing markets, transport provision and social reproduction. Setting aside abstract critiques of neo-classical economics or individualistic sociology, we stressed that the commonly held notion of the 'market' was not the best model for accounting for the complex interactions that we outlined. The twin problems of structured markets and interdependent submarkets made the broad idea of structured networks or institutions more meaningful. We pointed to further complexities due to the 'lumpiness' of housing, jobs and transport infra-structures, and the fact that they are spatially concentrated. Finally, we have outlined the changing structure of both economic activity and social life. The latter included growth in multi-earner households, growing female par-ticipation in the workforce, part-time working and a trend towards flexible, time-shifted and extended hours (in short, a move away from the 9–5 day). Associated with these were changing retail and leisure participation patterns (by location and time).

As everyone knows, choosing the 'best' location for home and/or the right job is not easy. Compromises have to be made, trade-offs considered. Where, as is the case for most of us, we are situated in multi-dependency households and/or local dependency networks, the decision is fearsomely difficult. Reflection on these problems, and the transport needs that they generate, suggests that there are a whole range of 'non-rational' or unacknowledged travel demands being generated in households and communities. Generally speaking, it is unlikely that adding a few more buses, or creating road pricing schemes, or providing low-cost travel or low-cost social housing is going to resolve these issues, or that any resolutions will extend beyond a select group in society.

Solutions to such complex issues are already in place. People living in cities solve the problem each and every day. They do so often at considerable personal social and economic cost. However, we have an opportunity to learn from people's everyday lives, and to use this evidence to reflect upon how the stresses and strains might be minimised. While we have established the value and importance of this exercise, any discussion of policy is premature. First, we aim to take account of, and be sensitive to, both the big and the little stories of life in the city. Having done this, we will try to establish a broad and meaningful conceptual framework within which we can detail and examine the different dimensions of our secret lives in the city.

We begin in Chapter 2 with laying out the conceptual and theoretical issues that we feel need to be addressed in order to provide an adequate foundation to study this problem. The outcome of this is a firm resolve to focus attention on social reproduction and the household. This also yields a clear argument for detailed ethnographic-style information collection about the daily lives of people in cities. In Chapter 3 we take the example of London and demonstrate the mismatch between the institutions that deliver, separately and jointly, movement, work, recreation and residence and the ways that people use them. This reinforces the notion of 'working against the grain of the city'. Chapter 4 follows up our central idea in more detail by looking solely at the household as an analytical lens; Chapter 5 extends this argument by discussing the complexity of strategies and tactics that people use to get by in the city. Chapter 6 returns us to some of the earlier policy themes about sustainable cities. The argument re-evaluates the urban sustainability debate and points to a missing concern with social reproduction issues. These are highlighted as the key points at which change could be effected. Thus far, existing policy initiatives have not addressed them. Finally, in Chapter 7, we elaborate the broader significance of the arguments developed in this book.

Chapter 2

Being there

In this place that is a palimpsest, subjectivity is already linked to the absence that structures it as existence and makes it 'be there', *Dasein*. But as we have seen, this being-there acts only in spatial practices, that is, in *ways of moving into something different*. (de Certeau 1984: 109; emphasis in original)

Introduction

The title of this chapter might, at first reading, seem to be enigmatic, yet it stands as shorthand for the issues that we want to address here. The opening quotation from de Certeau provides a conceptual mooring for this. More loosely, 'being there' encapsulates notions of everyday life and existence, practice and embodiment in space and time. As time-geographers have also commented, physical presence necessarily requires absence everywhere else: a person cannot be in more than one place at once. Being there signals a physical, embodied experience that is also an achievement because it implies prior communication, a decision to do one thing and to be somewhere.

We take the de Certeau quotation to be suggestive of three themes, each of which we try to elaborate in this chapter. First, the individual is not represented by a point, but rather should be understood as an intersection; individuals are constituted through their relationships to others. Second, the process of action, the 'in process', the 'doing', is an important perspective to counter that of the completed action, which is so often the focus of analyses. This alerts us to the possibilities of different ways of doing, or being. Third, being occurs in context. The city is not only morphology, nor is it just people. It is about how people use the city. Again, we would stress that cities and people actively constitute the lived experience that is the city, rather than the planner's vision. Significantly, people commonly find ways of being in cities that were either not expected or desired by their planners. While there are limits, people are endlessly creative about being in cities.

In Chapter 1 we outlined the trials and tribulations of ordinary people engaged in the effort of 'connecting'. It is a task that requires real struggle and compromise with colleagues, friends, family and strangers. It is situated both socially and economically, as well as spatially and temporally. It is this sense of how we get to 'be there' that we develop in this chapter. We outline a framework within which the processes and problems of 'being there' can be understood. This is our formal understanding of the 'secret life' of cities: the nitty-gritty of everyday life, which finds little correspondence in master plans, policies and much social theory of the city. It contrasts with the idealised, externalised public life of cities that we argue is the concern of most theories of the city. In order to sketch out our ideas we review and critique a number of attempts to navigate the dual attractions of individual demand and structural constraint. This chapter is a radical attempt to find a compromise between these two positions. Our theory allows us to offer a new framework for understanding the practice of living in and moving about cities. The notion is also implicit in the more prosaic desire captured in the title of Chapter 1: Only connect.

This desire to capture both individual and collective, structure and agency, is an enduring theme in social theory. It is our argument that insufficient attention has been paid to resolving these problems in urban theory. Structuration theory offers a well-known resolution and hence a potential way of thinking about cities, although it has not really been applied to a thoroughgoing urban analysis before. We take up structuration theory as a starting point for a critique that focuses on what we regard as the three central issues. First, the notion of institutions and networks. Second, the practice of everyday life, which we explore in its situated and embodied form. Third, a sensitivity to the constitutive nature of space–time and human action. This approach frames the research presented in the rest of the book. We seek to account for the practical and embodied dimensions of urban life instead of the more commonly apprehended idealised and formal elements.

The aim of this chapter is to develop a framework of ideas within which we can understand and further explore the complexities of urban life highlighted in Chapter 1. Our objective is to resist undue emphasis upon either voluntarist or structural formulations that either reduce cities to a myriad of individual choices or view them as a prison of power and control. We will stress the holistic nature of urban life that validates the interpenetration of economic, social and cultural life. Underpinning this is our concern with movement and embodiment, what Nigel Thrift (1997), echoing Pred (1977), calls the 'choreo graphy of the city'. Moreover, we seek to fold back these notions into core ideas of the social sciences, such as decision-making. In this case, we aim to resist instantaneous, rationalistic or mechanistic accounts of decision-making by opening up attention to the collective, ongoing, embodied and situated aspects of dilemma resolution.

We begin this chapter by reviewing attempts to develop partial, sectoral analyses of cities by looking at housing, employment, households and movement. The next section has two objectives. First, we open up our discussion to provide a critique of comprehensive analyses, paying particular attention to the notion of the 'dual city' that has been developed by Manuel Castells. Second, we critically explore key notions of network and institution as explanatory concepts. In conclusion we develop the constructive criticisms and reflect upon how an analysis of practice may transform the way we understand cities.

Fragments of the city: housing provision, employment, households and movement

The trajectory of much of the previous research on employment and housing relationships was touched upon in Chapter 1. In this chapter we are concerned with reconceptualisation of the relationships between housing, employment, social reproduction and movement. Allen and Hamnett's (1991) edited collection on housing and labour markets is notable for its argument that an analysis of the relationship between industry and housing (and transport) should be concerned with a conceptualisation both of the objects of interest (housing markets, labour markets and transport systems) and of their means of connection. Allen and Hamnett and their contributors make an important point about the significance of examining housing and labour, *not* as isolated elements, but through their connection. Furthermore, they stress the need to critically assess the nature of that connection: in practice this means problematising the adequacy of the 'market' as a concept.

Contributors to Allen and Hamnett's collection variously stress the complexity of the relations between employment and housing, the importance of class and gender, and both the historical and the spatial contingency of the actual outcomes of all of these relations. Two further points could be added to this putative research agenda: a consideration of the significance of transport and – more broadly – how the concept of 'institution' might be preferable to the rather fixed and unidimensional 'market'. Devising a wish list of topics for research is one thing; developing a clear conceptualisation that will justify and frame subsequent research is quite another. Hence, an important step is the exploration of the nature of the three elements – industry, transport and housing – and, coincidentally, their relation to one another via the household and social reproduction generally. Before moving on to such a reconceptualisation it is useful to clarify the more widely used concepts.

Housing provision

A key strand in the debate about the relationship between housing and labour markets at the regional level can be traced back to a paper by Thorns (1982). In this paper Thorns suggests that the structure of residential property markets reinforces regional labour market disadvantages. He argued that this process had, in the UK, given rise to a polarisation of the northern regions (unemployment, low house prices) and the southern regions (employment, high house prices). The simple notion that house price differentiation is a good indicator of the relationship between housing and labour markets has been a common one (see Thorns 1982; Coombes, Champion and Munro 1991). In a later paper Hamnett (1984), while agreeing with the causal mechanisms of this process, disputed the empirical fact that it had occurred at a regional scale, suggesting instead that the intra-urban scale might be far more significant (see also Barlow 1990).

Hamnett went on to argue that the idea of socio-tenurial polarisation (the socio-economic composition and income distribution of households by tenure) on a spatial basis may be more relevant than house prices as a causal process. Such a process, it had been argued earlier by Murie and Forrest (1980), might be amplified or attenuated by the unique historical characteristics of the housing structure of particular urban areas. The classic spatial form that this interaction of processes takes in Western European and North American cities is the polarisation of cities into inner and outer districts. Interestingly, work by Sharp et al. (1988) suggests that the empirical spatial pattern of such polarisation can be even more complex; in an analysis of Greater London they suggest that the traditional view of a doughnut of 'inner' versus 'outer' district categorisations actually obscures more significant differences that cut across such simplistic spatial patterns.

The relationship between socio-economic groups and housing tenure has long been a concern in the field of housing research. Hence, the basic subdivision of housing tenures has been developed alongside an appreciation of the nature of housing allocations into the concept of 'housing classes' (see Rex and Moore 1967) and urban managerialism (see Pahl 1969). This work was later criticised and reformulated by Williams (1982) to stress the institutional and organisational settings in which managers operated, thereby giving rein to analyses of housing set in a wider social and political context as well as admitting a concern with housing supply (housing production) rather than simply demand (see Ball 1983).

Three observations can be made about this body of work. First, the impulse of much of the research into housing classes and urban managerialism was to respond to the perceived differential access to housing by those located in structurally similar positions in the labour market. Second, the focus was primarily on the individual. Third, whether looked at from the supply or demand perspective, a salient feature of housing markets is their asymmetry. Only latterly has attention turned towards labour market factors, such as occupation, as well as the importance of asymmetrical relationships between supply and demand for housing and their spatially discontinuous nature (see Hamnett and Randolph 1986; Barlow 1990; Randolph 1991; Barlow and Savage 1991).

Employment

As the housing literature readily accepts, access to housing is clearly related to income and participation in the labour market (though not necessarily in a simple, one-way, causal manner). However, the analysis of labour markets proper does suggest that labour market participation is a complex process. Simple concepts of 'continuous' labour markets have been replaced by various 'discontinuous' forms – most notably dual and segmented labour markets. Continuous labour market theories characterise workers as if they were inter-changeable; they can do any job at the right price. Discontinuous theories of labour markets accept the notion that there are multiple labour markets, for example separate labour markets for chefs, and others for software engineers.

There are two main types of labour market discontinuity: spatial and structural. Spatial discontinuities can range from the designation of microlocal labour markets to global labour markets. Structural discontinuities can focus on skill/training, demarcation, race or gender. The most complex version is segmented labour market theory (see Krekel 1980). Research on the spatial elements of labour market segmentation has sought to interweave with an appreciation of structural constraints in order to argue that a combination of spatial and structural discontinuities create overlapping spatial patterns of structural discontinuity (see Cooke 1983; Peck 1989; Haughton 1990). The consequence is that apparently simple changes in labour demand (as a result of economic restructuring) or supply (through, for example, migration or retraining) at any one point in space or time will be experienced in very complex ways (see Johnson, Salt and Wood 1974; Salt 1991).

The work of Massey (1984) is indicative of the character of the relationship implied here. Massey's notion of the spatial division of labour incorporates the process of organisational change in firms and their employment consequences, as well as their spatial outcomes. Moreover, she also stresses that the historical existence of this process in any particular place may create the particular conditions of a labour market (which could be extended to include a variety of other local institutions that shape housing provision and movement) that might attract or dissuade firms to relocate. It may be argued that any particular locality is thus the outcome of a labour market structuration process. While Massey's work focused primarily on the regional and national scale, it could also be applied at the urban level.

Conceptualizing the relationship between housing and labour markets is clearly a complex problem. In an important and path-breaking essay, Randolph (1991) highlights the relatively autonomous and non-synchronous nature of labour and housing markets. A key concept that Randolph deploys is that of segmentation. His conception is as follows: the demand of firms for labour is mediated via labour market segmentation to produce spatially discontinuous labour market segments which vary between localities and over time; consequently, the position of individuals in the hierarchy of segments is a major determinant of their ability to consume housing. However, the past and current structures of housing provision must also be considered, as they constitute the historically existing stock; tenure plays an important role in structuring this segmentation.

Households

Importantly, Randolph also introduces the notion of using the household as a basic unit in such analyses: households are constituted by various combinations of individual labour market members and potential members. The social aggregation that is the household is not a simple one, as it may be modified by economic, social and cultural relationships within the household itself. Hence, while the infinite variety of households bears some relation to the position in the labour markets of individual members, it is far from being a simple or direct one. The key element here is the idea of the household, which is not elaborated in Randolph's work. It could be argued that the household can be viewed as a network or as an institution that mediates, or translates, social action (see Pratt 1996a). A household is a semi-permanent form of relationships: it may, empirically, have a wide variety of forms (not necessarily a 'family'); however, it is commonly delimited by the single housing unit. This form of social organisation is an important causal factor when the individuals comprising any single housing unit take into account other members' desires or needs. These internal relationships of household constitution are (causally) more important than the contingent relationships with the local neighbourhood.

While it is true that research on commuting patterns has highlighted the issue of the household in terms of demographic and socio-economic factors (Davanzo 1981; Camstra 1996) and the social environment (see Harbinson 1981), a strong causal model has not been developed. The notion of the household has been more salient in migration research (see Mincer 1978). However, usage in this work is usually reduced to the mediation of individual decision-making that ignores the issue of power as articulated via gender. Within housing research, the significance of the household as opposed to the individual as a unit of resource has been stressed (see Pahl 1984, 1988).

However, the work on household 'decision-making' or 'behaviour' is commonly modelled in a rationalistic fashion.

Against such assumptions is ranged a substantial body of work that stresses the diversity of the division of labour and resources (from both formal and informal labour) within the household. For example, Morris's (1988, 1991) work stresses that the structure of household dynamics is influenced as much by factors operating at the national level (such as the benefit system) as the ideological level (such as the patriarchal ordering of work and household relations) (see also the work of Grieco *et al.* 1989; Watson 1991). Moreover, work by Pahl (1988) on local survival strategies is also suggestive of a rather more extended notion of the household. Here, the concept of a local, self-serving and codependent community – engaged in extended social reproduction – could be used to capture the situated actions of individuals.

Movement

A weakness with much of the employment–housing analysis outlined above is with regard to transport. For example, Randolph concludes that labour and housing markets are linked through the 'local labour market' as defined by the commuting propensity of the various segments. This view can be criticised because it constitutes a voluntarist treatment of transport. This point is particularly serious as transport has been traditionally conceptualised through the dimensions of self (the individual), space (points of origin or destination) and mode. Connectivity has been considered indirectly through separate network analyses. It is notable that most transportation analyses are atomistic in their conception; they are not so concerned with processes but rather effects. The primary concern is with origins and destinations and auditing the flows between them. It is not surprising that this is translated into – or reduced to – land-use models rather than institutional or socio-structural analyses.

Transportation systems – by linking home and work – can, potentially, overcome or at least minimise the effects of structural discontinuities of housing and labour market segmentation by erasing the effects of space. An important strand of transportation research has long been concerned with the problem of the journey to work (see Lawton 1963, 1968, 1977; Warnes 1972; Thomas 1973; Dasgupta, Frost and Spence 1980; O'Connor 1980; Vickerman 1984; Simpson 1987; Cevero 1989; Giuliano and Small 1993). However, it has mainly been framed as a trip generation and allocation problem. Generally, aggregate demands of particular spatial units (either origin or destination) have been considered rather than the detailed composition of that demand.

Generally, we can note that this research on the journey to work, despite its considerable mathematical sophistication, has been carried out in the descriptive mode. This is not surprising given the considerable technical difficulties involved in the work, not least of which is the availability of appropriate survey data. Furthermore, researchers have tended to ignore, or considerably simplify, their modelling of the operation of housing markets; where they have considered labour markets they have been treated in a rather monolithic manner. Given the arguments outlined above concerning the relationship of housing and labour markets, one might expect considerable differences in transport needs for different segment groups – women, part-time workers, etc. – in order to overcome, or manage, labour and housing market discontinuities. *Inter alia*, this highlights the important historic role that transport has played in bringing together segments of the labour market, making it possible for individuals to utilise residential accommodation in one locality and employment opportunities in another. As such it could be said to naturalise or disguise the inefficiencies of both the spatial and temporal organisation of cities. In effect, it might be said that transport is the 'fix' or 'sticking plaster' over the cracks within and between labour and housing markets. As such it disguises much of their inefficiency and ineffectiveness.

Clearly, the issues of transport cost, time and system connectivity have to be considered. A point of view supported by recent research indicates that cost minimisation provides a poor explanation of commuting (Punpuing 1993; Giuliano and Small 1993). Another way of looking at this issue is to consider the fact that if costs are differently experienced by workers variously situated in segments of housing and labour markets, costs are likely to be more important for some groups than others. For example, more affluent social groups are able to resolve their particular housing and employment mismatches by the use of individualistic transport solutions (see Gordon and Molho 1985). Of course, in the case of the upwardly mobile commuter it may be an active choice to separate work and home (in the city and the country cottage) and, in such cases, the transport infrastructure and network availability may enable this.

Research on the emergence of long-distance commuting has touched on these issues. For example, Leyshon *et al.* (1990) explored the relationship between economic restructuring in the city of London and along the M4 corridor and related travel and housing patterns. They highlighted an issue of social polarisation and the settlement of particular localities by particular socio-economic groups. Housing choice, as a *consumption* choice, was possible due to the high incomes earned in particular occupations. Issues of career progression, using transport to gain access to spatially and structurally segmented labour markets, have also been highlighted by Fielding (1992).

The 'reverse side' of the facilitating nature of industry, transport and housing provision concerns the unemployed or those that are forced to use transport to gain access to employment. The unemployed may be in low-cost housing that is not well connected in terms of space, time or cost to the location of appropriate work opportunities. This is the spectre of the 'housing trap' that haunts many inner-city residents and, as such, this is an important dimension of the explanation of inner-city unemployment (examples for London can be found in Department of the Environment 1977; Vickerman 1984; Gordon and Molho 1985; Meadows *et al.* 1988). Individual solutions such as those summed up by the former Conservative minister Norman Tebbit for the unemployed to 'get on their bike' (that is, to move to where work is) may be effective for some, but it is collectively inefficient, as well as practically impossible, for some groups where cost, distance or transport network mismatches rule it out, or social limitations, such as dependants' needs, have to be taken into account.

It is not only the unemployed that should concern us here. Those in employment may be forced to make greater use of transport infrastructure in order to resolve their household and employment situations. As we noted in the previous chapter, this is not a new problem nor an exclusively urban one; people living in rural areas have traditionally had higher levels of car ownership than their urban counterparts due to the paucity of public transport. In some cases the cost of car ownership imposes an additional 'tax' on going to work. Current debates about road-use charging as a way of reducing congestion and pollution can also be criticised for having an adverse impact upon the 'private transport dependant' poor.

The dual city: towards an integrated analysis?

The analyses that have been outlined so far offer a number of insights, but they also suffer from a too broad brush approach that pays little attention to the interdependencies of movement, housing, employment and social reproduction. In this section we consider one contemporary holistic framework of urban dynamics, that of the 'dual city' which has been popularised in the writings of Mollenkopf and Castells (1991; Castells 1989, 1996, 1997). After outlining the key characteristics of the concept, we point to some of the underlying problems of using this analysis to help us to understand the secret life of cities.

The dual city and two missing middles

Castells's notion of the dual city has been elaborated and developed in a number of articles and is based upon a considerable body of evidence. The core idea is based upon two sets of interlinked processes: first, the absolute decline, or relative out movement from the city, of manufacturing industry; second, the disengagement of the state in the collective provision of services. Taking the first process in more detail, whether this is wholesale closure or relocation to cheaper premises (in the regions or in other countries), the result for the city is that manufacturing jobs have been lost in urban areas (see Fothergill and Gudgin 1982; Martin and Rowthorn 1986). The new jobs that have emerged are in different sectors and rely upon different skills. In spite of retraining efforts, it is seldom possible to re-place more than a small proportion of employees; those that are re-employed are commonly on lower wages, and less secure conditions. Commonly, jobs are found in low-level servicing jobs: cleaners, security, etc. So, the changes in the manufacturing industry are both structural and spatial. These changes are also paralleled by growth in the service sector (see Pratt 1994a for the London example).

The second set of processes is the 'rolling back of the state'; that is, the practice of governments, influenced by neo-liberal economic agendas (see McLennan *et al.* 1984), to aspire to minimal state expenditure and minimum economic expenditure and public debt (see, for example, Osborne and Gaebler 1992). In practice, such processes have been characterised by the contracting out of public-sector services to the private sector and, more generally, the removal of paternalistic and socially supportive employment contracts, and have been reinforced by the undermining of unionisation and the delegitimisation of union representation, plus massive attacks by the state on unionism. At the same time, employers have sought to capitalise on the flexibility afforded by such deregulation and introduced far more flexible and part-time working, as well as casual and contract working (see, for example, Lash and Urry 1987). Universal public service commitments have either been withdrawn or diluted in favour of either quasi-market forms or a two-tier provision where public and private providers work side by side. However, in practice, the quality of service delivery is quite different in each subsector.

In the seminal formulation of the dual city, Castells argues that these socio-economic processes interlock to produce two types of outcome: the structural and the spatial. The structural processes are of social and economic polarisation – the rich getting richer and the poor getting poorer. This phenomenon, evidence of which Castells and other writers (such as Wilson 1987, 1996) find in North American cities in particular, he terms the 'missing middle' or the 'hourglass society'.

The North American experience suggests that a substitution effect is occurring whereby manufacturing jobs are lost (the missing middle) and low-level service jobs are created to support the increasingly highly paid professionals. In global cities these low-level jobs are, more often than not, filled by economic migrants. Sassen (1991) has suggested that many other cities may also share the same trends. However, Buck (1994), in a commentary on Sassen's thesis, points out that the effects of income polarisation may be mediated by local welfare systems. Thus, in London, Hamnett (1996) argues that the same processes may not necessarily produce the army of low-paid workers, but instead a large group of unemployed.

The spatial dimensions of the dual city argument are even less clear. Castells describes a mosaic of spatialities that result from this process: the net result is a confusing juxtaposition of rich and poor. Castells and others (see Davis 1990; Soja 2000) have documented the rise of urban ghettos and enclaves that are a part of this transition. This process is shown *in extremis* in the form of gated communities in Los Angeles; enclaves that are quite literally walled and gated from the surrounding urban area. Gated communities are not unique to Los Angeles, nor to the current time period. Less sensationalist, but just as significant, are the banal and everyday processes of gentrification associated with tenure, income and cultural sensibility (see Smith 1996). Once again, it must be noted that different forms of housing provision and labour market regulation account for a range of structural and spatial outcomes in different nation states. While the spotlight has landed on housing and labour markets, movement and social reproduction have been overshadowed. The only notable exceptions are the cases of 'off the edge cities' that Soja (2000) refers to (see Chapter 1).

Castells's main argument is based upon a structural model of the city, albeit one undergoing a form of disintegration. The 'dual' in Castells's argument is the duality of social and economic networks. Castells's core notion, developed in both the 'information society' and the 'network society', is the 'space of flows'. In Castells's writings, the dual city is the single, physically delimited space where at least two spaces of flow intersect. For example, the global flow, which has London as a node, facilitates the transfer and exchange of know-ledge, capital, goods and people. At the same time, the local flow, such as to a travel to work area in the context of labour markets, limits the process to London or its region. In this way globalisation impacts upon the 'normal' expansion of an urban travel to work area through a smaller number of international migrants. As we noted above, the impacts are experienced in fragmented, segmented spaces and parts of institutions. Jobs, homes and social reproduction are aligned to one or other of these institutions.

Castells's dual city is mediated and structured in different ways due to the regulatory peculiarities of particular economies and societies. While he is able to provide a 'big picture', his analysis is less focused in terms of individual experiences. Despite Castells's evident concern (1997: Ch. 4) with social re-production, identity and nationality, his examination of the changing consti-tution of the household rests on issues of individual identity and structure and not on the reciprocal (or dependent) social, cultural and economic relations. Hence, the 'missing middle' of Castells's theory is how the spaces and flows are mediated (see Soja 2000: 212 *et seq.*, especially 229). Of course, Castells is not unique in having a weakness in his formulation in this area; this problem is a classic dilemma in social science. The classic attempt at resolution is Giddens's structuration theory. We turn to examine the potential contribu-tion of structuration in the following section.

From institutions to networks

In the last section we pointed out that the central problem lay not so much in the field of holistic concepts, that is, of linking employment, housing, transport, and social reproduction; rather, the problem was how to articulate them to account for, and describe, processes and how to get inside, and not to objectify, social action. Both sets of concerns intersect in the ongoing debates about structuration theory. We begin this section with a brief outline of struc-turation theory and use this as an opportunity to highlight the weaknesses of institutional approaches to analyses of social practice.

Structuration theory attempts to resolve an old social theoretical problem, that of the two perspectives of social analysis – structure or agency – and present them not as binary opposites but as a duality. A duality is a reformula-tion not reducible to either structure or agency (see Giddens 1984). While this is a deceptively simple proposition to make at a theoretical level, it is a difficult idea to develop in practice.

A key notion in Giddens's work is the idea of the 'duality of structure'. It is important to be clear about the usage of the term 'structure'. Giddens does not align his term to lay usage, for example as a building, or to traditional sociological usage, which refers to deterministic relations of the microscale (people) by macrostructures. Instead, he argues for a far looser terminology whereby structures are to be seen as the routinised interactions of people. Significantly, he emphasises that structures should not be seen only as constraining, but as enabling too.

In his formulation, structures are created through repetition and routine. The apparent stability of structures is just that: they are temporary and they can change. In order to capture the idea of duality, Giddens introduces the term 'institution'. Institutions are mediators, positioned between agents and structures, but they are reducible to neither. Empirically, the concept describes the organisations, both formal and informal, that we establish to regulate our societies. This is an uncontroversial sentiment, and the idea of looking at institutions is not novel. However, for us the import of Giddens's idea is captured in the sentiment of duality and the practical use of institutions as a lens on, or a way of seeing, human interaction.

Giddens's work has primarily expounded the notion of structuration in relation to class and social structure. However, we would argue that it can easily be applied to labour markets. Labour markets can be seen as the outcome of interactions between suppliers of labour (individuals differentiated by skill, education, training, experience, location and mobility) and consumers of labour (organisations, public and private, differentiated by needs for particular skills, experience, training and location). As anyone who has been present at a job interview will know, the interaction is not simply a rational process of matching characteristics of supply and demand; it is the result of compromise and modification. The particular characteristics of an individual may, in practice, redefine a job. This process, when repeated throughout many organisations, gives rise to an organic change in the structure of different segments of labour markets, which may in turn change the nature of 'demand'.

Giddens draws upon a body of work on time-geography, especially the work of Hägerstrand, to develop his notion of distanciation: the stretching of social interactions across space and time. Hägerstrand's pioneering work argued that time, while objectively the same everywhere, is not experienced, valued, used or available in the same way to all. In no small part, this is because time, for any individual, is also 'spaced' (see Hägerstrand 1975; Parkes and Thrift 1980; Gregory and Urry 1985). It is physically impossible to be in more than one place at one time, and there are absolute limitations on moving from one place to another in a finite amount of time. Initial appropriations of Hägerstrand's work stressed the novel forms of constraint upon the use of our time when coupling, or linking up with, others. These constraints are not simply those of the duration of particular activities – such as work, child care or shopping – but they are also linked to when and where those activities take place. Thus, particular combinations of activities are only possible for particular individuals due to their (that is the activities' and the individual's) time–space configurations (see Tivers 1985, 1988; McDowell 1989).

It is the spacing of social interaction that has drawn most comment in social science, especially when it has been applied to the notion of institution. Institutions also have a spatial dimension; their spatial extent – their 'footprint' – may be referred to as a locale. The footprint of an institution can be local or extend to the global. Building upon this formulation, geographers have argued that it is the combination of institutions that is particularly significant (see Gregory and Urry 1985). In summary, the concept of an institution does not refer to bodies that are simply concerned with 'education', 'parliament', etc., as in common parlance. Institutions can include these, as well as bodies such as local authorities or planning departments; they also may include non-formally constituted institutions such as households, labour markets and housing markets.

Institutions: third way or cul-de-sac?

Giddens is not the only writer to use the term institution; institutions play an important role in much recent theory that seeks to link the micro- and the macroscale. In the fields of sociology, politics, geography and economics, debates exploring approaches that stress 'neither markets nor hierarchies', or 'neither under- nor oversocialised accounts', and embeddedness have deployed institutional analyses (see Granovetter 1985; Hodgson 1988, 1993; Powell and Dimaggio 1991; Mingione 1991; Amin and Thrift 1994). As Hall and Taylor (1996) note in their synthesis and comparison of institutional approaches across the social sciences, there are numerous configurations of the concept ranged along a continuum from individual to structural. Another key analyst of institutions, Zucker (see Zucker 1988), highlights a key problem with this literature in the detailed analysis of institutions and change. She argues that theorisations are caught in the bind of either being insufficiently open to change, and hence appearing to be structural, or to facilitate so much change that they dissolve into voluntarism. Thus, while institutional analysis has cast light upon many new areas of social life, the search for a middle, or third, way seems to have stalled.

Practice: a social constructivist approach

Recent work on the embodied nature of human experience perhaps offers a new take on the resolution of the problem of mediation. Ironically, this intervention is based on a reading of time-geography, a field that Giddens draws upon in his formulation of structuration theory. More recent writing has drawn upon Hägerstrand's work in a slightly different way in order to highlight the situated interdependence of life and the pragmatic sense of possibility in practical situations of coping, or 'going on' (see Pile and Thrift 1995: 26). Such an approach strongly implies that the totality of outcomes is anything but the sum of total intentions, thus not totally predetermined by agents or structures, but rather worked out in practice (see Carlstein 1982: 61). Pile and Thrift (1995) suggest that such concerns link to other new writing about the flow of practice and embodiment. Specifically, they align this emphasis upon the pragmatics – the problem of resolving problems *in situ* – with a post-structuralist critique of sociology, such as that outlined by Game (1991). Game seeks to create an oppositional writing that works against what she claims to be the abstract and objectifying gaze of much conventional social theory, as exemplified by the work of Giddens. Game specifically seeks to write a social constructivist account of social action. Her account aims to recover the subjective, situated and embodied nature of social action.

In order to construct her argument, Game draws upon the work of de Certeau (1984) in her analysis of spatial practice. To appreciate de Certeau's approach it is useful to note that his understanding of movement is borrowed from, or echoes, post-structuralist accounts of the analysis of discourse and rhetoric. The point here, exemplified in Rorty's (1989) work, is the ways in which notions of truth are constructed through language. More technically, 'truths' are established by convergence, not by correspondence tests. Thus, the word 'table' and the thing referred to assume correspondence; convergence theorists argue that we collectively form agreements on what a table is. In short, our accounts of table converge through discussion and debate but, crucially, without reference to the essential 'truth' of 'tableness'. De Certeau draws upon a similar pragmatist approach to stress the dialogical construction of truth and understanding. According to such accounts we only know things under our various descriptions of them, not by their 'essence'.

How does de Certeau apply these notions to the city? There are two key points of relevance. First, he is keen to discard the essentialist notions of city (which he terms 'concept city') and he seeks to replace these accounts with one rooted in practices. The implication is that we should come to know our cities not through their external concepts, but in and through the practices by which cities are (contradictorily) constituted. Thus, a public space is not *a priori* a public space by virtue of its essential design characteristics of shape or form, but rather it emerges as one through public use and practice.

Second, de Certeau stresses that we should see disagreements and practical resolutions through the eyes of, or in the shape of, people's movement through the city. This dialogical and argumentative character of movement in the city is not random motion; rather it is purposeful and directed. It is a practical mode of finding what works, when and where. We are all familiar with finding the best route in a city. Again, it is not simply the rational, shortest distance; it is the embodied practice that we draw upon to substantiate such a statement. Thus, it is clear that movement is situated within a web of social relations of people, times and places. In short, actions can be seen as tactical and resolutions are achieved by individual action; in de Certeau's terms this is in practice, that is, 'on the hoof'. Although only suggestive, such a conceptualisation of urban life is one that places a focus first and foremost on the *practice* and the joint and ongoing nature of activities, as well as their embodiment and situatedness. We would argue that such a conception is an appropriate starting point from which to enter into an understanding of urban life.

This argument only takes us so far. Most analysts would allow for multiple ways of achieving an objective. However, they would be less willing to accept

challenges to the notion that capacity for movement, or resources, are a limit-
ing or enabling factor to rationally responding to demand. We would argue
against such a position, which is a crude reading of social action.

There is a danger in reading de Certeau only in terms of physical practices
and movement. Through extension of the argument we can quickly graft two
more insights drawn from discourse analytic work. First, and most obviously,
that it is also linguistic. Pratt (1991, 1996b) has shown how locality and
language are dialogically constituted and have practical effects. That is the way
that we describe and talk about cities, through plans and representations,
timetables and signposts. If we take a more explicit note of de Certeau we can
weave together the physical practices and the discursive realm and arrive at a
more satisfactory account of space and place. An important point is that such
account is not universally nor democratically agreed, nor is it an average of
movements or knowledges or accounts of them (a sort of travel to work area).
Rather, place, knowledge and practice are multiply constituted and lived
through. One possibility is dialogically linked to another.

A second, allied point that we can draw from the discourse analytic tradition concerns the discursive constitution of identity. A specific point that we
might consider here is that of decision-making. At the core of most existing
social, economic, geographical and political analyses is a psychological notion
of a unitary self. Such a representation normalises unitary attitudes and opinions, as well as the possibility of their rational expression and resolution. The
classic example is of 'revealed preferences', or, simply, demand.

Social psychologists working with the discourse analytic tradition have ques-
tioned the representation of attitudes as monolithic, unitary or immediately
accessible. Potter and Wetherell (1987) argue that identity is an intersection
of a variety of discourses; these discourses are selectively drawn upon by
people to justify, and account for, particular actions. Hence, any stretch of
discourse, say in an interview, may be expected to demonstrate contradiction
and variability. Analyses should, appropriately, be sensitive to such qualities.
Patently, traditional questionnaire analysis is not appropriate as it is based
upon a dualistic concept of language.

In summary, attitudes and opinions are socially constituted and dialogically
resolved as part of our situated, and ongoing, practice. This takes us several
steps beyond the traditional social theoretical foundations, and ways of seeing
cities. In effect, it turns them inside out by focusing on practice, not simply
seeing it as an outcome. It is also a substantial critique of traditional economic
accounts of knowledge and rationality. Such a conceptual movement opens
up a whole new realm to legitimate analysis, and exposes a greater proportion
of urban life to analysis in any given situation.

Conclusions: towards a secret life of cities

The aim of this chapter has been to provide a theoretical lens through which
the problem laid out in Chapter 1 would make sense and be seen as a real part
of the everyday life of cities. Our aspiration was threefold: to recover the

everyday life, commonly (although not exclusively) the social reproductive activities; the avoidance of idealisation, or reductivist accounts of the city; and to situate our analysis in the practices of people. These three elements we consider to be variously hidden, conceptually overlooked or excluded from conventional analyses and also policy formation debates.

In this chapter we have reviewed a range of theoretical work that impinges upon our concerns. We initially presented this in terms of the four motors that drive urban life. However, these accounts had shortcomings and the most serious of these was the focus on structural determination of social action, or on voluntarism. We saw little opportunity for understanding everyday life in its complexity here – in fact, the accounts seemed configured to overlook it. Policy-making in this field seemed to follow a similar course and, likewise, it has misdirected its attention to structural-technical fixes.

Our analysis in this chapter has sought to strive for an integrated analysis, rather than superficial integration after the fact that is so often the case in policy-making. We examined integrated urban analyses, such as those of Castells. However, they were found lacking, from our perspective, due to the weakness of their central mediating concepts between macro- and micro-processes. We looked to social theory, particularly that of structuration theory, to provide answers. While we had sympathy for its general aims, the application did not seem to offer a focus on the practice of urban life. However, a rich tradition of time-geography that runs through structuration theory, and informs post-structural sociology, was found to be a good foundation on which to build. We also found the work of de Certeau, on the dialogical nature of social action and the importance of the tactics of social action, to be the basis of a social constructivist account of urban life.

We have found the notion of institution, and its associated looser term 'network', to be useful in our analyses. We do accept that both terms are contested, and problematic. In our writing we seek to imbibe these terms with a new inflection. In short, we see networks and institutions as outcomes, and not as starting points; they are constituted through the practice of everyday life in the city.

The title of this chapter sought to capture our concern: the substantial achievement of 'carrying on' one's everyday life. 'Being there' also resonates with the theoretical framework that we have sought to develop with its focus on 'practice'. We draw our inspiration for the rest of the book by turning our attention to practice-based materialist analyses that seek to explore the dialogical nature of life in the city. The methodological implications that we draw from our approach are to focus on ethnographic-style accounts. In practice, we have had to collect this information through extended interviews; however, we have sought to reconstruct the flow of everyday life in our accounts. In so doing, we hope to have started a process of the recovery of knowledges about the dilemmas, problems and possibilities of getting by in the city.

We are under no illusions that getting by in the city is free, fair or equal. There are established and historical structures and established resistances of power and influence. As we noted in Chapter 1, and we have sought to touch upon here, these resistances are of a particular form – they are both material and human, and they are selective in their effect. In an earlier paper (Pratt 1996a), we sought to capture such an idea in the notion of the 'grain of the city'. This grain facilitates, more or less, a particular category of identities/ meetings/times/places.

Thus, the time-geography of Tivers (1985, 1988) could be said to chronicle attempts to work 'against the grain' of the city. The point is that the flow of 'the grain' emerges not by chance, but runs along particular lines. Tivers highlights the case of women with caring responsibilities for children; however, by extension, time-geographies could be constructed for others within the population. We would resist the notion of pointing to groups, or segments, *a priori*. Instead we seek to highlight the situated practices that may stretch across times and spaces and involve many other actors (human and not). Some of these practices are prioritised; they work with the normalised conception of the city. Other practices (commonly associated with marginalised groups, but not exclusively) have to work against the grain of the city. In some cases these people find it impossible to 'go on'.

We will argue in our final chapter that the banal practices of the secret life of cities are also the microsites where inequalities are reproduced – but, potentially, they are sites where inequalities could be challenged. However, before mounting this argument we need to present our evidence and sustain the case against the 'gaze' of the normalised city, the public life of cities, and begin to look for the secret lives that we all know so well.

Chapter 3

Making the connections: the dual life of a city

Introduction

This chapter addresses the relationship between global and local practices, focusing attention on the intersection of social reproduction, or everyday household life, and processes of modern urbanisation. In particular, it highlights the time–space coordination of different daily practices and what we term the institutional structures for households living in one large city – London. Issues like work, home, shopping and daily movement, as well as their corresponding institutional structures of employment, housing, retailing and transport, are used as the analytical 'point of entry' from which we explore key connections between the global and the local life of cities. As might be expected, the former constitute the most basic moments of everyday urban life and the latter have always been the foci of urban policies. We prefer to see cities not as separate islands, but rather as inextricably linked, one to another, by a web of production systems, finance and resource usage, and by the environmental problems they both create and face. More than ever before, urban residents, businesses and politicians are shaping the economy, society and environment of cities, and they are doing so in a profoundly global context. In short, recognition of this global–local interconnection – 'the local in the global' and 'the global in the local' – provides an essential dimension to understanding the secret life of cities in the twenty-first century.

London has been chosen as our case of study because of its scale and its leading roles in global, national and regional production and consumption. It demonstrates close links between the global and the local in both production and consumption spheres. London is large enough to facilitate collection of primary and secondary data, both as a city and with respect to contrasting neighbourhood study areas. As such it lends itself to comparative in-depth microlevel research. We believe that the approach used in this chapter could also be applied to smaller towns and cities, in different regions, in developing as well as advanced world economies.

This chapter begins from a more conventional standpoint by reviewing London's overall 'institutional structures' relating to employment, housing, retailing and transport. It then moves on to highlight the significance of the routinised practices of social reproduction in sustaining these institutions. It illustrates the bonding forces, as well as the microcontexts, of urban processes. The chapter concludes with a discussion of both the tensions that constitute the city – the city that many of us live in, and move about, every day – and the broader institutional context within which this activity takes place.

Defining London

In order to focus attention on both the global and the local life of cities, an urban definition is required that covers the major boundaries of productive and reproductive activities in terms of a *daily urban system*. On the one hand, this suggests a definition that combines regional and functional boundaries in a wide regional context. In practice, however, this inherently unstable urban definition poses significant practical problems with respect to collection of secondary data such that, while it is conceptually attractive as a definition, it is empirically impractical. On the other hand, it points to administrative and physical boundaries such as the London boroughs and Greater London. These provide the building blocks of most official statistics and the bases of resource allocation under current political systems. Although the green belt policy restricts expansion of the built-up areas, in fact economic growth and social change transgress these physical buffers and administrative boundaries. Unlike

Victorian and Edwardian London, where jobs, dwellings and shops were concentrated in a relatively small area, 'post-industrial London' in the twenty-first
century is moving towards wide-dispersed development. We are forced, therefore, to adopt a more flexible view of urban definition by setting the discussion in a wider regional context, such as to see London as an extended region
stretching into the south-east of England, or in a discrete context, such as
Inner London. Such a definition should ideally reflect the life lived by people
rather than that designated by administrators. In practice, the research draws
upon statistical data collected on administrative bases for the sake of data
availability and the consistency of comparable spatial units. In some sense this
reflects the conflicts between conceptual and practical definitions of cities and
the inappropriateness of current institutional arrangements of urban jurisdic-
tions. Accordingly, the changing boundaries of cities (in socio-economic terms)
are themselves the outcomes of cities' developmental processes.

London is here defined in terms of Inner and Outer London areas, reflecting the simplest and most obvious zoning (administrative) division. These
zones are further subdivided according to the physical division (by the River
Thames) of north and south London, and the socio-economic division of east
and west London. Figure 3.1 illustrates this 'quartered' city effect. When
viewed in association with the functional/regional view of extended London,
these internal divisions can be used as benchmarks to understand the spatial
features of London's institutional structures in relation to employment, housing,
retailing and transport.

Inner London[1] accounts for one-fifth of the area of Greater London
(approximately 200 square miles). It consists of 13 boroughs and the City of
London and represents the area equivalent to Victorian London lying immediately beyond the commercial core and the West End. By contrast, Outer

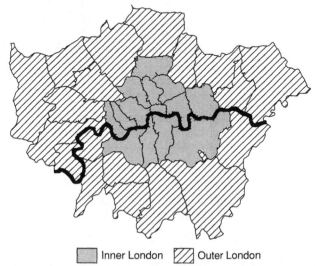

▨ Inner London ▨ Outer London

Figure 3.1(a)   London as a 'quartered' city: Inner and Outer London (Source: Hall, 1990)

West London   East London

Figure 3.1(b)   London as a 'quartered' city: East and West London (Source: Hall, 1990)

London's 19 boroughs cover an area of 782 square miles representing the more prosperous and suburban image of twentieth-century London (Hall 1990). The River Thames provides a significant physical north–south divide, despite the numerous bridges and tunnels which connect opposite sides. While the underground network is denser and extends further in north London, the area south of the river is mainly linked by surface trains. Moreover, the train stations in south London are further apart than their underground counterparts in north London. Living and travelling in north London and south London imply quite different experiences for both Londoners and visitors.

A strong case can also be made for dividing London according to an east–west axis. While no physical or administrative manifestation exists, many Londoners view this division as obvious and self-evident. It largely represents a complex interaction between geographical evolution, industrial development, political struggle, cultural change and social dynamics that jointly write the history of London. In effect, west London is closely related to growing industries and prosperous communities, and east London is linked to industrial and community decline, unemployment and social deprivation. By examining the intersections of the internal spatial categories of regionalisation (into Inner–Outer London, north–south London and east–west London) and the functional categories of regionalisation (into different daily locales), it is possible to highlight the time–space connections between different institutional structures in London.

Decentralisation and disparity: a world city divided

As a daily urban system, the expansion of London's boundaries and its merging with adjacent towns suggests a persistent trend of decentralisation. This is particularly evident in terms of employment (deindustrialisation), housing (population decentralisation), retail development and transport. Each of these institutional structures are introduced in turn below to highlight the different patterns, processes and contexts of decentralisation and disparity in these spheres.

Deindustrialisation: a combination of 'shake-out' and 'spill-over'

In 1991, London employed 3.5 million workers, accounting for 20 per cent of the total employed population in Great Britain (Office of Population Censuses and Surveys, OPCS, 1994a).[2] This is compared to a population of 6.7 million, representing just 12 per cent of Great Britain's total population (OPCS 1994b). While London functioned as a major workplace in the region, it was experiencing a significant decline in numbers of workers with some sectors losing more than a quarter of a million jobs between 1981 and 1991. Viewing London as an extended region, however, the picture is one of substantial employment gains in the part of south-east England resting outside Greater London (the 'Rest of the South-East', or 'RoSE') and a corresponding decline in Inner and Outer London. This suggests employment decentralisation in the metropolitan area. But this process did not occur evenly in every sector. There was both a restructuring of employment and a redistribution of job opportunities. Generally speaking, manufacturing employment declined substantially over the years – nearly half of the manufacturing jobs in Greater London had gone in the ten years between 1981 and 1991, a loss of more than 350,000 jobs. Service employment, by contrast, had a 3 per cent increase in the same period (Department of Employment, *Employment Gazette* 1983–93). This suggests that there had been a continuing shift of London's employment structure from manufacturing to service sectors, a process which was described as the *deindustrialisation* of London's employment and industry (see, for example, Martin and Rowthorn 1986; Sassen 1991; Graham and Spence 1995).

Deindustrialisation was characterised by a combination of the 'shake-out' of manufacturing jobs and the 'spill-over' of service jobs in London. On the one hand, manufacturing employment declined substantially in the south-east region as a whole, but the extent of decline was more pronounced in Inner London boroughs than in Outer London and the RoSE, that is, the 'shake-out' of manufacturing jobs. On the other hand, service employment grew considerably in both Greater London and the RoSE, but with higher rates of growth in the RoSE while Central and Inner London maintained a leading role in higher-order services. This represented the 'spill-over' of service jobs.

While there is not scope here to examine the various factors underpinning London's employment restructuring, it is important to recognise key interactions between global economic restructuring, the creation of new intra-national spatial divisions of labour and the historical geography of local industries (see Massey 1984; Barlow and Savage 1987; Castells 1989; Harvey 1989a; Ernste and Meier 1992; Dunning 1994; Conti *et al.* 1995). Together these resulted in a serious mismatch of job opportunities and local labour markets. In the 1980s, for example, about half the available job vacancies in Greater London, including a majority of unskilled and retail/catering vacancies, were located in Outer London, while the majority of managerial, professional and clerical vacancies were concentrated in Inner London (Meadows *et al.* 1988). Many of these downtown white-collar jobs were taken by those who lived in the suburbs; but many Inner London workers, especially those who were looking for retail, catering and unskilled jobs, found that they were more likely to find suitable work in the outer rings of London. These labour market mismatches have profound economic, social and environmental implications, such as reducing London's overall economic competitiveness, resulting in fractional and structural unemployment, ultimately creating a growing need to travel a longer distance to work.

Housing market segmentation

As might be expected, the inefficiency of London's labour market, necessitating the import of labour from areas outside London and the creation of skill gaps between London's job opportunities and its workforce, has much to do with London's housing structure. Housing market segmentation is particularly evident with respect to tenure. The uneven development of private owner-occupier, private-rented and social-rented housing is such that private-sector provision is dominant in Outer/west London and social housing heavily concentrated in the Inner/east zones. While private housing represents 70 per cent of Greater London's total housing stock (LRC 1993), only 9 per cent of Inner London households lived in owner-occupied accommodation in 1991. The stark tenure divide between east London boroughs, for instance, where social housing accounts for the lion's share of local housing provision (Tower Hamlets 75%; Southwark 64%; Hackney 61%; and Islington 61%) (LRC 1993), and Outer/west London boroughs, where owner-occupation is the norm, is partly explained by the history of London's industrial development (see Greater London Council, GLC, 1975; Wohl 1977; Slater 1980, 1981).

Some 20 to 30 years ago, London had a more balanced employment structure between manufacturing and service sectors. Moreover, London's housing and employment structures used to be very close to each other in space and time. Jobs in textile factories, furniture workshops, breweries, ship-building docks, warehouses and markets were more readily available in the capital city, especially in the inner areas of east London. In these areas, inexpensive social-rented housing fulfilled the housing needs for a great number of London's workforce, including immigrants from the north and overseas. Likewise, luxury apartments clustered around commercial centres in the West End and the City of London. However, with the extension of London's transport systems, people were able to move more efficiently between the workplaces in central areas and suburban housing developments and this accentuated the dispersal process.

The Greater London housing stock totalled 2.9 million in 1991, of which 40 per cent was concentrated in Inner London and the remainder distributed at generally low densities in Outer London (LRC 1993). Since new dwellings may take years to complete and account for a very small proportion of existing housing stock, London's housing structure remains remarkably similar today. This rigidity in housing development contributes to the strong outward dispersal of London's housing and population, a trend which is more marked still if the RoSE is taken into account. While London's employment and industrial structures have changed dramatically as a consequence of global, national and regional economic restructuring, the housing market is particularly slow to accommodate this change. Indeed, London has had the lowest rate of new house-building in the UK over recent years.[3] There are, therefore, both *quantitative* and *qualitative* mismatches between London's housing and employment structures. In Inner and Central London, jobs outnumber dwellings. In the outer rings of London, dwellings outnumber jobs. In short, the adjustment mechanism of the housing structure in London is both slow and lacks flexibility.

The consequences of structural mismatches are, among other things, the growth of 'wasteful journeys' in the Greater London region and the problem of concentrated unemployment in some Inner London boroughs. The problem is that the cost of a 'mobile society' is very high in terms of economic efficiency, social equity and environmental quality. Moreover, pressures towards greater dispersal of employment, housing, transport and retail development pose particular constraints for those social groups that do not have equal degrees of mobility, notably the poor, the unemployed, the disabled, the elderly, women and children. On the one hand, outward moves undertaken essentially for housing and environmental reasons have a major effect on dispersing labour supply; at least for those workers able to make such moves. As Buck *et al.* (1986: 46) argued back in the 1980s, 'there was little sign that gentrification has significantly reduced the net outflow of professional or managerial workers from Inner London.' On the other hand, because such moves are closely related to income, the population of Inner London included an increasing proportion of those who could secure only poorly paid jobs and those who are effectively trapped in housing provided by local authorities and housing associations.

It is not a new idea to view housing and employment as an integrated issue; they are related to each other in many perspectives, such as the intimate relationships between labour supply and the housing market (Allen and Hamnett 1991), the coordination between child care and paid employment for women with young children (Tivers 1985), and the time–space tensions between urban structures and household lives (Pratt 1996a). Nevertheless, what is missing in existing discussions is a holistic view that sees individual actions and reproduced structures as one integrated issue. This is made particularly apparent when London's transport systems and retail developments are also taken into account.

Transport: bridge or barrier?

Having said that to bridge the institutional disparities between London's housing and employment structures people and products/services have to be increasingly mobile, movement has, in effect, become an essential part of urban life. It is not surprising that when we talk about modern human 'settlement', very often we have to deal with the problems of 'unsettlement' first. We therefore have to examine issues of transportation. The scales and patterns of different kinds of trips made in London illustrate the significance of transport in linking residential and employment locations. On a typical weekday in 1991, over 20 million trips were made either wholly or partly within the Greater London area between 07:00 and 21:00 (LRC and DoT 1993). Apart from 4 million trips which were made entirely on foot, there were 16.6 million 'non-walk' trips of which almost 10 million constituted motorised trips made by car, either as drivers or as passengers. A quarter of motorised trips were made for work purposes and were typically concentrated in two peak hours in the morning and the evening. Looking at the overall pattern of commuting, it is evident that one in five of those who worked in Greater London travelled from outside; and this figure rose to one in two in Inner London (OPCS 1994c). As might be expected, the growing scales of motorised trips in terms of their numbers and distances pose significant challenges for London's transport systems and for people who either work or live in London. Consequently, questions concerning the transport linkage between mismatched employment and housing markets need to be set against the physical opportunities and constraints presented by London's current transport structure.

Broadly speaking, London's transport structure is characterised by a dichotomy between public and private transport. Public transport, especially the underground system, is more readily available in Inner London and for inward commuting (including some nodal locations along the main lines of rail and underground networks), in particular in areas north of the River Thames. Private transport, by contrast, is more typical in the outer rings of London and for outward commuting. In 1991, for example, public transport served nearly half the work-related journeys for the locally employed workers in Inner London, while private car accounted for less than a quarter of local commuting trips. For those who worked in Inner London but lived outside, public transport served 68 per cent of the inward commuting trips; car trips accounted for just 28 per cent (OPCS 1994c). In Outer London, by contrast, only 17 per cent of those who both lived and worked in the area used public transport, but as high as 55 per cent of them used private cars. For those who worked in Outer London but lived elsewhere, only 18 per cent of them used public transport, but as high as 78 per cent of them drove to work (*ibid.*).

In many respects, then, London's transport structure has reinforced the structural disparities between employment and housing. Decentralisation of both population and job opportunities has increased the number and distance of journeys in the extended London region. This suggests that, given current pressures of crowding and congestion, the prospect is growing inequality between mobile and trapped sectors of the population. At the same time, however, London's transport structure reduces 'time–space friction' between London's employment and housing structures, regardless of whether these work-related trips are made by public or private transport. In doing so, it serves to maintain the city's leading role in high-order services by importing skilled and professional workers from London's hinterlands to meet the labour demand in Central and Inner London.

In summary, the growth of longer and orbital journeys, increased crowding on roads and railways (trains and, in particular, the underground), and higher rates of unemployment in some Inner/east London communities, all suggest that London's transport systems are insufficient to accommodate the growing disparities between jobs and housing. Growth in motorised journeys in general, and routinised journeys to and from work in particular, have created enormous threats to the environment in terms of air pollution, noise and the waste of fossil energy. Equally, lack of mobility has become a major obstacle to some social groups gaining access to jobs and housing, as well as other resources. As might be expected, it is disadvantaged groups, such as the poor, the unskilled, women and part-time workers, that are most seriously affected by the push for a more mobile society (see Focas 1985; GLC 1985; Grieco et al. 1989; Labour Party 1991). Accordingly, the 'transport issue' cannot be discussed in isolation, as an issue of mobility per se. Rather, it should be discussed in association with other institutional structures, as an integrated issue of accessibility to resources and opportunities (see Glaister 1991; London Planning Advisory Committee, LPAC, 1996).

Retail development: changing regimes

Cities are in many ways the meeting ground for the stretched time–space practices of global production systems and the routinised acts of daily urban consumption. The concentration of buyers and sellers and the functions of circulation, exchange and consumption constitute an important part of the urban landscape. Moreover, after work-related trips, shopping trips represent the second largest category of daily movement in London (LRC and DoT 1993). As Bromley and Thomas (1993: 2) note:

> The contemporary city is to a substantial degree articulated in relation to retail facilities, and this has important consequences for the nature of city growth and associated opportunities and constraints for urban planning . . . since the vast majority of the population is involved in some direct or indirect way with shopping activities.

While a top priority of recent urban policy has been to reduce the need to travel by private car (see DoE/DoT 1994; DoE 1996b), retail development has largely functioned to subvert this goal. In London this is partly a feature of a polycentric retail network. Unlike many smaller cities and towns, London is not a single place. Generally speaking, it consists of a loose network of at least

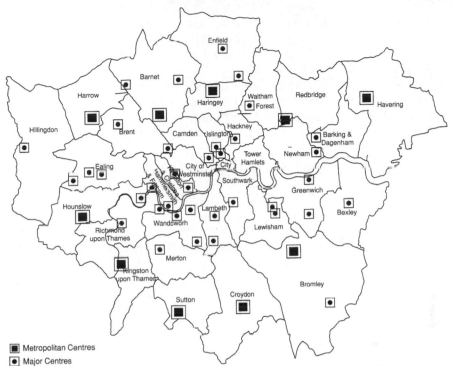

Figure 3.2 London's polycentric retail network – a hierarchy of retail development
(Source: URBED, 1994)

five categories of retail centre: one *international/national centre* in Central London; 10 *metropolitan centres* forming an approximate ring around Outer London; 38 *major centres* more or less evenly distributed in Greater London; at least 150 *district centres*; and 32 *out-of-town shopping centres* (URBED *et al.* 1994). This polycentric network and its resulting hierarchy of retail development are illustrated in Figure 3.2.

Three major changes in retail development can be observed to feed into this polycentric feature. First, changes in ownership patterns have led to the growth of large national and international corporate retailers and the growing dominance of multiple chains at the expense of smaller, independent and more local operations. Second, there has been a persistent trend of retail decentralisation. Third, there is evidence that 'retail gentrification' is also taking place in the capital city.

The 1970s and 1980s saw dramatic change in British retailing, characterised by some as a 'retail revolution' sweeping through the country (Bianchini *et al.* 1988; Worpole 1992). A notable trend in this 'revolution' has been the growing concentration of commercial power in the hands of a relatively small number of firms and the continuing growth of large multiple chains (O'Brien and Harris 1991). In the 30 years between 1960 and 1989, for example, the multiple chains increased their share of the total national retail sales from

33 per cent to 80 per cent (Worpole 1992: 18). This dominance registered most significantly in bulky household shopping and, in particular, grocery shopping. Wrigley (1993: 41) estimates that the market share of the top five grocery retailers increased from under 30 per cent of the national sales in 1982 to 61 per cent in 1990. This trend is as marked in London as in the rest of the UK. The net result has been a decrease in the overall number of shops and growth in the average size of retail floorspace for individual stores. This has generally been at the expense of smaller, independent and local concerns with little power to compete with multiple chains (Wrigley 1993; Burke and Shackleton 1996).

A second trend is the decentralisation of London's retail development. This is characterised by a continuing, though slow, decline of traditional town centres and a fast growth of out-of-town shopping facilities, such as superstores, hypermarkets, retail warehouses, retail parks and out-of-town shopping centres (Westlake and Dagleish 1990). This does not mean that the high streets and in-town shopping facilities have disappeared altogether, but their roles in shopping provision have changed substantially. Unlike other metropolitan cities or small towns, London's traditional town centres, in particular the international/national centre in the West End of Central London, still retain a fairly strong position in comparison shopping and leisure facilities. This is due to the concentration of different kinds of shops and transport networks that have established London's privileged position with respect to retail provision. In other words, what has happened to the decentralisation of London's retail development is not a simple trade-off between in-town and out-of-town locations. Rather, it is a more sophisticated pattern of retail differentiation characterised by the decentralisation of some retailing facilities, notably grocery and bulky goods and out-of-town shopping, and the continuing concentration of comparison and leisure shopping in central locations.

Finally, despite widespread dispersal of larger retail outlets to off-centre locations, there are at the same time good examples of *retail gentrification* in London. This represents a competing trend whereby specialist and 'high end' retail units are moving back to central locations. Covent Garden is the best-known example (see O'Brien and Harris 1991). With the exception of some downtown specialty shops, however, retail gentrification is most marked in grocery retailing. After nearly two decades of decentralisation and floorspace expansion, some of the leading supermarket chains have begun to open smaller stores in central locations, such as Central London and the high streets. For example, Tesco successfully opened their new 'Metro' stores[4] in Oxford Circus, Covent Garden, Goodge Street and other high-street locations. Apart from serving the shopping needs of the shrinking downtown population, these gentrified supermarkets tend to focus on the 'transit population', such as downtown employees, metropolitan-area shoppers, inner-city residents and tourists. In fact, some large food multiples like Marks & Spencer have been very successful in appealing to downtown workers by providing more expensive, high-quality prepared and ready-to-eat food in smaller packages. Moreover, new office and commercial developments now typically incorporate some

Photograph 3.1 Bringing food shopping back to the city (East London Tesco): space for car parking remains (Source: Helen Jarvis)

combination of retailing facilities. Tesco and other small shops at Canary Wharf in Docklands are good examples of retail gentrification (see Photograph 3.1).

In summary, then, the dominance of multiple chains, the decentralisation of bulk retail provision, and the gentrification of some grocery retailing, all suggest that retail development by and large parallels associated changes in London's employment, housing and transport structures. The decentralisation of population to suburban areas within the extended London region represents a decentralisation of 'effective demand' that has important knock-on effects for London's retail structure. Moreover, the mobility of suburban households has increased substantially with the growth of car ownership and this has altered the shopping habits of more mobile and affluent London residents. While the decentralisation of London's retail structure has brought benefits of greater choice, lower prices and better shopping environments for more affluent consumers, it has at the same time raised significant new obstacles to less mobile, disadvantaged groups. More importantly perhaps, changes in retailing also have significant employment implications. In London, retailing employed 20 per cent of London's working population in 1991 (together with hotel and catering) (Department of Employment 1993). It also remains the most important source of employment for women and part-time workers (Townsend 1986; Sparks 1992; Watson 1992). Moreover, the rapid expansion of part-time (female) work in retail and the deregulation of shopping hours has been one response to women's rising participation in paid employment and reduced availability to shop during 'normal working hours' (Reeves

1996). In turn, changes in retail structure affect the time–space relationships between employment, housing and transport.

An integrated city?

As illustrated above, there are increasing disparities between institutional structures in relation to employment, housing, transport and retailing in London. Judged against the needs of most users, they don't mesh together. On the one hand, advances in transport technology and extension of transport networks have largely reduced the time–space friction between decentralised activities for some people, notably richer and more mobile groups. On the other hand, the increased need to travel in a 'mobile society' has created extra barriers to gaining access to resources and opportunities for others, especially women, the poor and other disadvantaged groups. It could be argued that a world city like London should be a city for all, be they rich or poor, residents or visitors, current or future generations. This requires an integrated approach to coordinate the mismatches of institutional structures.

First, institutional coordination has to overcome sectoral barriers *between* departmental agencies in the process of policy-making. Too often policies are made based on the projection of future demand characterised by trend planning within sectoral boundaries. This implies a need to create a constructive and productive dialogue between the full range of stakeholders in legislative and administrative spheres, leading to negotiation and cooperation platforms for decision-making at all relevant levels, as well as between public and private sectors.

Second, institutional coordination also requires integration *within* sectoral boundaries. No two individuals are totally the same. While stressing the need to establish a holistic framework for the coordination of institutional structures, one should bear in mind that different individual needs should not be carelessly averaged out or summed up. It means alternatives should be assured in order to coordinate different types of development in relation to employment, housing, retailing and transport. For example, different types of transport systems must be planned together as one integrated system. Only when intrasectoral aspects are fully taken into account in the process of policy-making is it possible to realise intersectoral coordination.

Most importantly, both *intra-* and *intersectoral coordination* should have proper regard to the quality of *internal coordination*, that is, institutional coordination based on intrinsic links between activities and associated time–space organisations in the practices of everyday life. Urban structures are not only the media that constrain and enable people's daily activities; they are also the very outcomes of people's actions and reactions. Unless the basic needs of coordinating different daily moments are properly addressed, it would be nonsense to talk about any institutional coordination. For example, when commenting on the planning implications for a sustainable city, Breheny (1993: 72) argues that 'there is little point in creating an alienated community for the sake of energy conservation from high densities.' This means that the internal dimension of institutional coordination cannot be reduced to nominal links and numerical parities of spatial integration.

Last but not least, given the decentralisation of both population and activities in the capital city, coordination of fragmented institutional structures may require the scale of analysis to be extended to a regional context. This would ensure greater scope for both intra- and intersectoral coordination.

Time–space coordination and social reproduction

Central to this chapter is the argument that it is the connections between routinised practices of everyday household life and institutionalised social structures that need to be addressed if we are to achieve a form of sustainable urban development. In this section we turn our attention to social reproduction: the other side of the coin. Neither rejecting structural constraints nor neglecting the significance of individual opportunities, the overall aim is to recognise the need to examine both extensive structure and intensive action. It is important to observe both the institutional and practical life of cities because the practices of everyday life effectively draw upon, generate and reshape the structural features of urban institutions and vice versa.

This section introduces qualitative evidence from in-depth household interviews as a means to get inside the intimate sphere of social reproduction. The aim is to explore the life-world of individual household members with reference to the ways they are embedded in both household relations and local spatial attachments. Unlike many microanalyses that unquestionably equate agents with individual persons, our analysis takes the *household* as the principal unit of empirical analysis. Households (including traditional families, single-parent families and unmarried couples) are basic exemplars of social institutions through which the intrinsic links between different institutional structures can be observed, in particular via the practices of coordinating different daily moments between household members. Many decisions in everyday life are made on the basis of household units rather than on the basis of individual persons, such as where to live, what to eat and whether to drive or not. By focusing on household life it is possible to vividly reveal the 'contextuality' of 'institutional webs' and 'networks'. This refers to a concept describing institutional relations necessarily involving other sets of relations by virtue of the interrelated nature of household life. These complex webs provide the secret links between London's employment structures, housing markets, retail developments and transport systems. Accordingly, greater understanding of the significant interpersonal conflicts underpinning the organisation of different household members' everyday lives, as well as the compromises made by them in order to live together as a consensual group, highlights an important dimension for further consideration – that of institutional coordination in space and time.

We draw upon in-depth interviews with households living in two contrasting London boroughs: Tower Hamlets, located in the Inner/east zone, and Harrow, located in the Outer/west zone (see Figure 3.3). These locations

Figure 3.3 Location of Harrow and Tower Hamlets study areas in London

represent contrast between inner-city and suburban London structures. Each study area includes two separate neighbourhoods selected to reflect local variance in terms of the socio-economic backgrounds of sample households and the structural properties of local environments. The two sample areas employed in the Tower Hamlets study are Bethnal Green (a traditional East End community) and Wapping (a gentrified residential area in the Urban Development Area of London's Docklands). The Harrow study focuses specifically on Greenhill (a residential area with a good mix of terraced houses, converted flats and semi-detached properties near Harrow town centre) and Stanmore (a more prosperous area dominated by semi-detached houses). Contact was made with more than 400 households and 40 households were finally selected to proceed to interview stage. Those finally selected represented a wide range of life-course stages and commensurate time–space configurations. A technique of semi-structured, in-depth interviewing was used, with all the adults in the selected households interviewed separately. Table 3.1 provides a summary of the households interviewed in each of the four neighbourhoods. In each case, pseudonyms are assumed for the interviewees. For the sake of convenience, couples are referred to as 'Mr' and 'Mrs' to denote gender. Surnames starting with 'G' refer to Greenhill households, with 'S' referring to Stanmore households, 'W' Wapping and 'B' Bethnal Green respectively. Details of the research design and method are included in the appendix.

Generally speaking, there are two kinds of household decisions that are relevant to the coordination of everyday life. First, there are 'material household decisions', such as taking or changing a job, moving home, getting married, giving birth to children and so on. These decisions are made relatively infrequently, perhaps just a few times in a lifetime. But by virtue of their impact and permanence, these material decisions shape the time–space configurations of a particular household's daily life in profound ways. Second,

Table 3.1 Profile of London households interviewed, organised according to four study areas:

(a) Greenhill, Harrow

Interviewee Household	Household Composition[1]	Car Ownership	Employment Status[2]	Locations of Employment	Mode of Transport
Gill	H,W 3C(7,11,14)	1	H(FT) W(X)	Paddington —	Underground
Good	H,W 1C(12)	1	H(FT) W(X)	Harrow —	Walk
Gordon	H,W	2	H(FT) W(FT)	Hounslow Kilburn	Drive Drive
Glen	H,W 4C(2,4,4,6)	1	H(FT) W(X)	Waterloo —	Underground
Graeme	SM 2C(11,13)	1	SM(FT)	South Harrow	Drive
Guy	H,W,E 1C(7)	1	H(FT) W(FT)	North London Harrow	Underground Drive
Gadd	H,W 3C(3,4,6)	1	H(FT) W(X)	Old Bailey —	Underground
Gerton	H,W 2C(13,16)	2	H(FT) W(PT)	Harrow Harrow	Walk Walk
Game	H,W 2C(7,11)	1	H(FT) W(FT)	Wembley Oxford Street	Drive Underground

[1] Household Composition: H — Husband; W — Wife; SM — Single Mother; E — Elderly; C — Children
[2] Employment Status: FT — Full-time; PT — Part-time; X — Not Working

(b) Stanmore, Harrow

Interviewee Household	Household Composition	Car Ownership	Employment Status	Locations of Employment	Mode of Transport
Simon	H,W	1	H(VJ)[1] W(VJ)	Harrow Harrow	Drive Drive
Seal	H1,W1;H2,W2 1C(1)	2	H1(X),W1(FT) H2(FT),W2(X)	Hounslow City	Bus Drive/U-ground
Smith	H,W	1	H(FT) W(X)	S. Kensington —	Drive
Star	H,W 3C(7,12,13)	2	H(FT) W(FT)	Harrow Kenton	Drive Drive
Slade	H,W 1C(6)	2	H(FT) W(FT)	Ealing Kingsbury	Drive Drive
Shaw	H,W 1C(8)	2	H(FT) W(X)	Harrow —	Drive
Sharp	SM 2C(12,14)	1	SM(X)	—	
Sullivan	H,W 2C(10,12)	2	H(FT) W(PT)	Guildford Wembley	Drive Drive
Sand	H,W 2C(10,14)	1	H(X) W(FT)	— Camden	U-ground/Drive
Spiller	H,W 2C(10,13)	1	H(FT) W(PT)	Park Royal Harrow	Bus/U-ground Drive

[1] VJ — Voluntary Work

Table 3.1 (cont'd)

(c) Wapping, Tower Hamlets

Interviewee Household	Household Composition	Car Ownership	Employment Status	Locations of Employment	Mode of Transport
Wade	SM	1	SM(FT)	Isle of Dogs	Drive
	1C(16)				
White	MP,FP[1]	2	MP(FT)	City	Walk
			FP(FT)	City	Walk
Wallis	H,W	1	H(X)	—	
	2C(2,11)		W(FT)	Home	Drive
Weaver	H,W	3[2]	H(FT)	Oxford Circus	Drive
	2C(2,5)		W(FT)	Oxford Circus	Drive
Wexford	H,W	2	H(FT)	Tottenham C. Rd	Drive
	1C(3)		W(FT)	Battersea	Drive
Wilton	MS[3]	1	MS(FT)	Maple Cross	Drive
Warde	SM,MS	0	SM(FTS)[4]	South Bank	Bus/U-ground
	1C(8)		MS(FTS)	Hammersmith	Bus/U-ground
West	H,W	1	H(FT)	Kent	Drive
	1C(6)		W(X)	—	
Winter	H,W	1	H(FT)	City	Drive
			W(FT)	Knightsbridge	Bus/U-ground

[1] MP — Male Partner; FP — Female Partner
[2] One company, one household car, and one antique car (for display)
[3] MS — Male, Single
[4] FTS — Full-time Student

(d) Bethnal Green, Tower Hamlets

Interviewee Household	Household Composition	Car Ownership	Employment Status	Locations of Employment	Mode of Transport
Bevan	H,W	1	H(FT)	Bethnal Green	Walk
	2C(7,11)		W(PT)	Bethnal Green	Walk
Berry	H,W	1	H(FT)	City	Walk/Bus
	2C(16,19)		W(X)	—	
Burton	H,W	1	H(FT)	Islington	Drive
	1C(24)		W(FT)	City	Bus
			C(X)	—	
Baxter	H,W	0	H(FT)	City	Bus
	2C(17,19)		W(FT)	Shoreditch	Walk
Barker	H,W	3	H(FT)	Potters Bar	Drive
	2C(21,24)		W(PT)	Shoreditch	Walk
			C1(FT)	Poplar	Drive
			C2(FT)	Shoreditch	Walk/Drive
Broom	H,W	1MB[1]	H(FT)	City	Ride M-bike
	1C(23)		W(PT)	Bethnal Green	Walk
			C(FT)	Hackney	Bus
Boyle	H,W	0	H(FT)	Victoria	Underground
			W(FT)	Chancery Lane	Underground
Bayer	H,W	1	H(FT)	Bethnal Green	Drive[2]
			W(FT)	Islington	Bus
Bliss	MS,E	0	MS(FT)	Piccadilly	Underground
Blunt	SM	0	SM(FT)	Croydon	Train
	1C(7)				

[1] The husband rides a motorbike
[2] The husband is a minicab driver

there are 'routinised daily practices', such as commuting decisions, daily or regular shopping activities, children's schooling trips and the associated escorting trips made by parents. Because these routinised practices are repeated day to day, they are typically dismissed as being mundane or trivial. While routine practices might indeed seem trivial for each household individually, cumulatively and collectively, routinised daily practices associated with travel to work mode, or management of the school run, have a significant impact. Significant social and economic costs of congestion, for instance, and negative environmental impact can be traced back to large numbers of individual households each coordinating complex home–work–family spatial relations with recourse to the private car.

These two household decision types are closely related. At the same time, however, reinforcement of one by another varies between households according to circumstance. To some extent, then, the distinction made between the 'material household decisions' and the 'routinised daily practices' echoes Giddens's argument for a contrast between 'discursive consciousness' and 'practical consciousness' (see Giddens 1984: Ch. 2). We find that this distinction is a key to a deeper understanding of the social nature of day-to-day life. The remainder of this chapter draws out these institutional connections with reference to individual and household experience of coordinating material decisions and routinised practices in relation to employment, housing, shopping and transport.

As the largest city in the UK and the nation's most import centre in politics, economy and culture, London provides jobs for more than 3.5 million people. It is the base for a quarter of a million businesses, including three-quarters of Britain's financial and business services (DoE 1993b). Nevertheless, the shape and profile of London's labour market has changed dramatically over the last 30 years. As we noted above, the processes of deindustrialisation resulted in the shake-out of manufacturing and the spill-over of service-sector employment opportunities. It was also explained how the inherent unevenness of these combined trends contributed to severe mismatch in the distribution of jobs and housing in terms of both quantity and quality. Not only is it the case that jobs and houses may not be available in any proximate fashion, necessitating that workers live and work in different areas, but few households actually 'rationalise' home and work locations to minimise travel to work as a first priority, especially when viewed with respect to change over the life-course. This is especially the case in London where the competition in land use and job markets is most severe and transport networks are relatively good. In these circumstances, living in the centre of London may not necessarily mean being central to the job market and living in suburban London may make sense as a central location for both London and the RoSE when this is viewed as a regional rather than a place-specific market.

Changes of workplace and changes of the home–work relation

The time–space dynamics between home and work can be illustrated by the case of Mr Baxter, who grew up in the East End community. Mr Baxter was

made redundant from a local brewery, after 25 years of continuous employment, when the firm shut down in the early 1980s. Since then he has tried
several different jobs, either working locally or working elsewhere in the
London region. For him, the process of deindustrialisation in the capital city
makes the 'central' location remote to job opportunities for a man with his
manual skills.

In contrast, London continues to offer the greatest concentration of job
opportunities in the UK for professional and managerial workers. For example,
Mr Wilton (single, mid thirties) is a computer engineer providing database
services to companies on a contract basis. He typically finds the majority of
his short-term job opportunities in Central London:

> London is a very big job market, in particular the specialised job market. It is
> probably the biggest job market in the UK that we can commute within. It makes
> it feasible for me to be able to get contracts somewhere within travelling distance.

In order to live close to the job market, Mr Wilton bought a two-bed flat in
Wapping several years ago. At the time he made this purchase, he was working
in the City. Since then he has changed job several times, taking on new contracts,
but each new workplace has been within a 20-mile radius of Central London.

It is widely recognised that job turnover is higher in London than in other
cities (Savage and Fielding 1989; Fielding 1992). Frequent job change can
pose particular problems for individual workers according to occupation, household employment structure (households relocating more than one earner)
and with respect to high transaction costs associated with owner-occupied
housing relocation. Job change can be especially problematic for those employed
in specialist occupations and senior positions. Mr Spiller, for example, was
recently made redundant by a pharmaceutical firm where he had been employed as a senior analyst for six years. He explains: 'The professional considerations are far more important than travel considerations . . . [because] you
are not always able to find what you want within your desired location.'
Relocation may be feasible for younger and/or single people but for those
who have family considerations or have a financial or emotional attachment to
a particular house and location, increased travel to work might be preferable
to relocation. Moreover, the problems of coordinating home and workplace
locations are particularly problematic for dual-earner households, and even
more so if these represent competing careers in specialist occupations.

It is also unrealistic to talk about fixed workplace locations for occupations
such as architecture, civil engineering, construction, landscaping and so on
where place of work shifts according to changing project requirements rather
than firm location. For example, working as an architect, Mrs Sand has been
employed from a Camden-based firm since 1990. Her first project was a joint
programme in partnership with another company in Tottenham Court Road.
At the earlier stages of design work, she spent most of her time in the Tottenham
Court Road office, staying there for three years. At later stages, she was asked
to move into the construction site in Stevenage, some 30 miles away from

London. However, before this particular project was completed, Mrs Sand was assigned other projects. One project entailed design work in the Camden office, with visits to the construction site in Portsmouth. Another project had an engineering office in New Malden, South London, but the construction site was in Sandwich, Kent. So in a typical week at the time of interviewing, Mrs Sand had to spend one day in Portsmouth, two days in New Malden and another two days in Camden. Occasionally she had to go to the construction sites at weekends. Clearly, the fixed spatial relationship between Mrs Sand's Stanmore home and her Camden office did not mean much to the organisation of her daily life.

Finally, Mr Gordon believes that this is 'an age without job security'. He draws on his experience of having to be mobile, changing jobs several times over the last few years as a local authority housing manager, to explain the disadvantages of assuming a fixed residential location. In his view, 'There is no point buying a house near your job [because] you don't know when you will change your job . . . [and] it is very difficult to get a job locally.' Most importantly, 'you've got all the mortgage, you can't move home easily . . . so we are prepared to travel'.

Suburban living, household life-course and first-time buyers

People choose particular residential locations for very different reasons. Journey to work considerations effectively contribute just one element of a complex equation. Given the dense networks of both public and private transport in London, households arguably have a wider choice between different residential locations. Many factors affect a particular household's housing decisions, such as locations, prices, property features, community amenity, environmental quality, public facilities, transportation links, schools, neighbours and the potential capital gains in local housing markets. While there is not scope here to explore all the various factors that affect housing choice and location, it is important to remember that housing decisions are of great long-run significance, especially where this involves the purchase of residential property. Housing is both an amenity and a positional good (Hamnett 1999). Most importantly, housing decisions are not 'one-off' affairs, but rather closely related to household life cycles (Rossi 1980; Clark and Onaka 1983; Kendig 1984; Forrest 1987).

Many suburban households interviewed in this study previously experienced Inner London living. As they advanced through the stages of family formation, these households typically sought access to a bigger house, more rooms, a large garden, safer and quieter streets without through-traffic, adequate local facilities, such as libraries and schools, and a better environment (see Photograph 3.2). For the growing family, houses are typically more desirable than flats or maisonettes, and concentrated residential areas better than flats above shops. In these circumstances the housing market effectively 'steers' working families towards the outer parts of London by the underlying

Photograph 3.2 'Metroland': expanding the horizons of Londoners' housing prospects, 1908 (Source: London's Transport Museum)

structure of housing tenure, quantity, type and quality. House prices have traditionally been significantly higher in London than the UK average (30 per cent higher in 1983 rising to 50 per cent higher in 1987 and stabilising at 35 per cent in 1994) (Government Office for London 1995). It is not easy for a first-time buyer to own a property in Outer London, not to mention in more expensive central locations. Indeed, for many 'traditional' households, moving to the outer suburbs essentially provides the most feasible way of combining family accommodation with metropolitan life. In effect, the search for better environmental quality and more affordable housing effectively contributes to greater home–work dislocation for this population of family households.

Children's schooling, investment and housing decisions

Children's education ranks high among the top priorities of many household location decisions. The need to be close to a 'good' school can be more important than the need to be close to one or more workplaces. Many parents want to send their children to 'good' primary and secondary schools. This typically means those schools performing well according to league tables based on GCSE results. Not every household can afford to send their children to private schools and many parents are ideologically committed to state schooling. In such circumstances, good local schools represent one of the key factors that define a good residential environment.

Generally speaking, the division between 'good' and 'bad' schools in London more or less coincides with the division between Outer/west London and Inner/east London. For example, schools in four Outer London boroughs (including Harrow) were listed towards the top of national GCSE league tables in 1995/6 (Department of Education and Employment 1997). In the same year, seven Inner London boroughs (including Tower Hamlets) were listed towards the bottom of the league tables. As Mrs Star points out, 'Children's education is the thing we think about all the time'. The majority of family households interviewed in Greenhill and Stanmore mentioned that the reputation of local schools played an important role in their housing decisions.

In Inner London, by contrast, many local schools' GCSE results fail to match the quality associated with the higher prices fetched by local residential property. In Wapping, for example, Mr and Mrs Wallis sent their 11-year-old daughter to a boarding school in Scotland for this reason. Mr Wallis explains: 'The local school was terrible . . . [pupils had] no homework, no discipline . . . Basically we don't believe in private education . . . but we've got no choice; we sent her to a private school.'

Similarly, Mr Weaver (father of two children aged 2 and 5) explains why he and his wife are contemplating a move from centrally located Wapping to the outer suburbs:

> Our house is one of the biggest in this area, and the location is fine. We've got the facilities here and we've got a fairly good environmental quality. Children can play on the back [yard], we can park our cars very easily, and travel to work is fine. So we are quite happy with the area. But one thing we are worrying about is education. The secondary schools here, on the whole, seem not terribly brilliant. The published league tables indicate that they are not very good. We don't know whether we can afford the private schools; and private schools are not necessarily better. We have to consider very seriously, when our children are growing older, whether to move away from here, primarily for educational reasons. The trouble is, to get the house with the size we want in Inner London, it's just too expensive. Therefore, we certainly have to move outside London, such as Hertfordshire or Buckinghamshire . . . [because] the education systems are better there.

This emphasises that home is not just a shelter. It represents a range of issues relating to the very reproduction of household life. Nevertheless, in discussing this dimension, it must not be forgotten that housing can contribute to household wealth in the realm of production. Some homeowners benefit from the significant build-up of positive equity between the purchase and subsequent sale of a house in a 'good' location. For example, redevelopment of London's Docklands has meant that Wapping is now within 'the largest redevelopment area in Western Europe and the greatest opportunity since the Fire of London' (Brownhill 1990: 1). As many interviewees admitted, this factor was playing an important role in their housing decisions.

Commuting: an issue of accessibility

Housing and employment dynamics vary from one household to another. They also change at different stages of household life cycles, as described above. It is difficult, if not impossible, to coordinate such diverse needs for different household members in a small area. As might be expected, some, or all, household members must make certain compromises if they want to live together as one household. Indeed, increasing numbers of 'two-dwelling' households are forced to live near the workplace during the week and go 'home' to live with other household members at weekends (Green, Hogarth and Shackleton 1999).

Household members frequently have to coordinate an increasingly fragmented life in a wider time–space zone. Very often, more and longer journeys are the answer. For policy-makers and environmentalists, ironically, the all-too-tedious journeys to work or shop have profound socio-economic and environmental implications, particularly given their repetitive nature. A small time–space friction between home and work for individual persons may generate a significant volume of movements for society as a whole. Arguably, then, it is insufficient to focus on patterns of movement *per se*, such as the distances, times, directions and modes of transport. More importantly, planners and policy-makers need to recognise the diverse local contexts and household dynamics behind the seemingly obvious transport patterns.

Difficulties in coordinating housing and employment decisions are evident in the case of the Seals, a family household comprising three generations. Mr Seal recently retired from his job in India and rejoined his family in London. Mrs Seal is employed in a Hounslow-based electronics factory, a job she has held for over 20 years. The couple's only son works for an accounting firm in the City; their daughter-in-law also lives with the couple as a full-time housewife, taking care of two children aged 1 and 3. The Seals moved to Stanmore from a rented flat in Hounslow. They wanted to buy their own home and considered several possible locations, including Hounslow, Richmond, Wembley and Harrow. They finally chose their present address in Stanmore because it offered affordable housing together with a preferred residential environment. However, 'in order to have a bit of both', Mrs Seal now has to change bus three times and spends 75 minutes every morning getting from Stanmore to her workplace in Hounslow.

At the one mobility extreme, a small proportion of London's population walk to work (less than 10 per cent in 1991: OPCS 1994b). For example, both Mr and Mrs White walk to work from their Wapping home to their office in the City of London (they work for the same firm), a walk journey of 15–20 minutes. In Bethnal Green, similarly, both Mr and Mrs Bevan go to work on foot. Mr Bevan works in a take-away shop, 5 minutes' walk away from their home, and Mrs Bevan works in a small textile factory just around the corner of their home. Nevertheless, the rationale underpinning these seemingly equivalent pedestrian journeys are very different. In Wapping, some households (like Mr and Mrs White) chose this location in order to live close to their workplace in Central London. In other words, relative housing mobility facilitated this consolidation of home and work. In Bethnal Green, by contrast, many short-walk journeys simply reflect the fact that those 'trapped' in poorer inner-city housing experience very limited access to job opportunities further afield. As Mrs Bevan said: 'I do this job because it is convenient, not because I like it . . . Home is more important than work.' The constraints of domestic responsibilities, the fixed nature of social-rented housing and a lack of suitable transport force many inner-city households to look for local job opportunities. Furthermore, it also suggests that a particular person's transport decisions are, to a certain degree, decided by their housing and employment decisions.

At the majority end of the scale, 90 per cent of London households are unable to walk to work and, inevitably, large numbers make recourse to the private car. London has among the worst road traffic congestion in the UK. At the same time, Central London offers the most extensive (and dense) network of public transport in the UK. The question why so many people choose private over public transport must be answered with reference to the complex interdependence of production, consumption and social reproduction activities and their interface with existing institutional structures. With the UK government calling for a modal shift from private car to public transport, it needs to be recognised that car trips accommodate significant temporal as well as spatial gaps which open up in people's daily lives. Public transport cannot adequately meet the scheduling requirements and spatial arrangements of increasingly complex and dislocated modes of urban living. Not every place in Inner London is well served by public transport, although it may be well 'covered' by some sort of public transport network. Shorter distances do not necessarily imply that travelling time can be largely reduced. According to statistics, access, waiting and changing times account for more than half the total door-to-door time for short radial rail journeys, and nearly three-quarters for central area journeys (DoT 1994: 200).

The inadequacy of London's public transport systems is most marked in the outward movements to, and the orbital movements around, the outer suburbs. For example, Mr Wilton, who lives in Wapping and works in Maple Cross, north-west London, has to drive over an hour every morning to work. He explains:

> The place I work at the moment is not really accessible by public transport. They [the company] run a mini-bus from the tube station. It is not convenient to take ... When my car goes to service, I do use public transport ... But it is not a pleasant journey ... and you have to take a taxi from the tube station to the office.

Paradoxically, a move from central Wapping to a more suburban location may not reduce car dependence.

In the case of Mrs Guy, a secretary in a local Greenhill school, it is suggested that travelling short distances by public transport is time consuming in suburban areas because of the absence of extensive and dense networks. Mrs Guy points out that

> Working locally is quite good. But working locally [in Outer London] without [private] transport is very difficult. I could not go to work by bus; it would take me an hour ... it is not practical ... If I drive, it takes only fifteen minutes.

Orbital movements around the outer suburbs are more problematic still for those who have to rely on public transport for their daily commuting. Mr Spiller, for instance, travels a relatively short distance each day from his Stanmore home to his workplace in Park Royal. To drive this distance would take no longer than 25 minutes, but travelling this route by public transport is particularly convoluted as Mr Spiller describes:

> It is a difficult journey for me from here [Stanmore]. London is very good for commuting if you want to go directly into the centre. If you want to go around the area, the orbital route is not so brilliant. It would be twenty-five minutes' car ride [to Park Royal], but it [public transport] takes me, on an extremely good day, at least fifty minutes. On a bad day, it could be an hour and a quarter to an hour and a half.
>
> I would walk from here [home] to Harrow and Wealdstone station. It takes me fifteen minutes or so. Then I would go four stops [on the Underground] to Wembley Central. From there I take a bus to Sudbury Town, and then I would travel on the Underground two stops to Park Royal. The office is three minutes' walk away from the station.
>
> Coming home, that is slightly faster. I go from Park Royal to South Harrow by tube [the Underground], and then I would go from South Harrow to Harrow Weald by bus. From there it is about seven to eight minutes' walk back home ... [The reason] why I do that in the evening is because Harrow is less congested at that time. If I try to do it in the morning, it just takes far too long because Harrow gets so congested in the morning. It is a combination of people going to work and children going to school. That's why I have to choose such a complicated route to work. That route is the quickest route, the optimal route, to the office by public transport.

Given the inconvenience and frustration of such journeys, it is not difficult to imagine why many people make recourse to the private car when their financial circumstances permit. This said, this argument is particular to orbital travel. Radial commuting, in contrast, offers greater potential for public transport commuting given London's current transport structure. For those who live near a station, a slightly longer-distance train journey to central locations is a very efficient way of commuting in terms of the total time of journey and the average cost per mile of travelling (Roberts 1992: 253; Banister 1994: 65).

Making the links between home, work and movement

In a sense, recognition of orbital and irregular movements in the outer suburbs represents a more realistic view that takes into account the scale of existing suburban developments in the metropolitan area. It suggests that a sustainable transport strategy for London should not rely on a single prescription of either more road building or rail extension; nor should it resort to a spatial integration that restricts the developments in small areas at the nodal locations. Rather, the issues of commuting and transportation must be understood as one part of a more fundamental issue, that of coordinating everyday household life. This holistic view suggests that we must analyse this issue in a wider household context instead of a narrowly defined question of individual locations, distances, frequencies and modes of transport.

For example, many people take their car to work either because they need the car for work purposes (making sales calls for instance) or because they use the car on the way to and from work (to escort children to and from school for instance). This critical 'balancing' required to coordinate everyday home, work and family life is illustrated in the case of the Slade household.

Mr and Mrs Slade live together with their only daughter who is six years old. Mr Slade is a chef, working in Ealing. Mrs Slade is a health visitor, working from a clinic based in Kingsbury. They moved from Harrow Weald to their current address in Stanmore two years ago, mainly because they wanted a bigger house and a better environment or, as they put it, 'a better quality of life'. Both Mr and Mrs Slade drive to work, partly because they need to travel back and forth between home, workplace and school during the day and partly because Mrs Slade has to use the car to visit patients. Mrs Slade explains why they feel they carry the high cost of running two cars as the best way of coordinating their daily routine:

> In this part of Stanmore, the [public] transport is not particularly good. There is only one bus . . . I use the car for my work because my job involves visiting people. That is why I need the car . . . and also because I have to take my daughter to school, which is about half a mile away. I need the car for that. My husband works in Ealing. In order for him to come home early enough without taking public transport, it is easier that he has a car too.

Mrs Slade went on to describe a typical day in their daily lives:

> I leave the house at about 8:40 in the morning, taking my daughter to school and leaving the school at 9:00. I then drive from Stanmore down to Kingsbury; it takes me about twenty minutes to reach the clinic. So I arrive at the clinic at about 9:20 . . . I probably leave the clinic [to visit the patients] at 10:00 and come back at 12:00 . . . I pay someone to pick up my daughter from school three days a week. But the other two days I normally do this: I rush from work at 3:00, running and picking her up from school and bringing her back home, then waiting for my husband to come back. He will be back by car at about 3:45. I then go back to

work for another hour's job and leave at about 5:30, coming home at 5:45 . . . But in the past five years, I was also working on a degree at college (Westminster University in Harrow). Some mornings I have to go to college, so I drop my daughter to school and rush to college, staying at college in the morning and then going straight back to work at about 12:45. Because I also have to go to college in the evening, when I come home at about 5:45, I will see my daughter and talk to my husband for a while, cooking and eating, and then go out to college again . . . and come home at about 10:00. I do this twice a week and sometimes three times a week.

The Slade's story suggests that the concentration of assorted developments at nodal or central locations may largely reduce the time–space frictions between different daily moments, and thus largely reduce the need to travel. Nevertheless, an enabling condition for one household may mean a constraining condition for another. This points to the need for *variety* and *flexibility* as the prerequisite for a sustainable transport policy that prioritises *accessibility* over *mobility*. These qualities can only be assured when the issues of commuting and transportation are discussed in a wider social context of coordinating everyday household life.

Shopping: the changing contexts of urban life

Finally, the interrelated nature and compromising character of everyday household life requires examination of household shopping practices. This is because shopping trips, apart from work trips, are the second-largest category of motorised trips in London, and the largest single category of trips made in off-peak hours (LRC and DoT 1993: 28). Moreover, the spatial manifestations of current retail development have profound implications for the cities (see, for example, Wrigley 1993; House of Commons, Environment Committee 1994; Raven and Lang 1995).

Two types of shopping activities need to be differentiated. The first is daily or weekly shopping for food and other groceries, or obligatory shopping. The second is discretionary shopping, or the occasional non-food shopping for comparison items or leisure purposes (Tivers 1985: 135). These two categories are not mutually exclusive. For some people, food shopping is as interesting as leisure shopping. For others, non-food shopping is a tedious chore. In many circumstances, one shopping trip is aimed to meet the needs of both food and non-food shopping. However, for the sake of simplification and to generate deeper understanding of the interrelationships between shopping activities and other aspects of household life, the following discussion concentrates on issues concerning food shopping.

As explained previously, shopping in London entails multiple choice: households usually have a range of choice between different types and locations of shops. In Wapping, for example, households have a choice between small local shops in Wapping Lane, a large local supermarket (Safeway) in Vaugham Way, non-local market stalls in Whitechapel, and non-local large supermarkets in Surrey Quays (Tesco, in a large shopping complex), the Isle of

Dogs (Asda and Tesco) and Whitechapel (Sainsbury's), not to mention other possible shops and locations in the capital city. Similarly, households in Stanmore have a wide range of choice between different directions and distances of travelling assuming widespread private car use for suburban shopping trips. Places like Wealdstone, Hatch End, Edgware, Harrow town centre, Colindale, and Brent Cross are the most frequently mentioned locations near the study areas.

Local food shopping: space matters

From the interviews it was suggested that, all things being equal, household members tend to use the nearest shops, regardless whether they live in Inner or Outer London, with or without a car. In Wapping, for example, households like to use a large local supermarket. They might also purchase smaller items from shops in Wapping Lane. Likewise, in Bethnal Green people tend to do their food shopping in Bethnal Green Road, where a medium-sized Tesco is situated alongside small shops and street stalls. In Greenhill, households tend to shop in a large Tesco (with a car park) at the edge of Harrow town centre. In all but one of the study areas (Stanmore), shopping facilities were within walking distance. Overall, the decentralisation of grocery retailing into the residential areas has substantially reduced the distances between residential and market places. For those without access to a car, this does to some extent reflect a lack of choice. For example, Mrs Wilton, who must rely on public transport as her only means of transport, explained:

> It is inconvenient to shop in other areas . . . because the [shopping] bags are so heavy and the bus is so small and crowded. Otherwise, you have to get a minicab, which is expensive. But the things are not so much different [between local shops and the shops in other areas], so I tend to use the [local] Safeway two or three times a week . . . usually on the way home.

More generally people tend to undertake short-distance local food shopping on foot, whenever possible, although many households have no difficulty gaining access to a car. In part this is due to the convenience created by the proximity between homes and shops and partly due to the difficulty of carrying heavy and bulky shopping bags. The tendency is to use local shops (including supermarkets) more frequently by buying fewer and smaller items. For example, Mrs Weaver typically buys food for a single meal from the local supermarket on her way home after picking up her daughter from school. In addition, she does a 'major shop' on a weekly basis to get bulky things like mineral water and cans.

Paid employment, gender role and food shopping: time–space matters

Despite significant changes in gender divisions of labour over the past 30 years, notably with regard to increased female participation in paid employment, repeated evidence from 'time–budget' surveys confirm that women undertake most food shopping and other daily activities of social reproduction (Bowlby 1987; Devault 1991). Nevertheless, the average time spent on shopping has reduced, especially for those in paid employment. Working women typically spend 2.93 hours per week on food shopping, one-third less than non-working women (the *Independent* 24 March 1995). Together with the popularity of time- and labour-saving home appliances, larger storage space for both fresh and readily prepared food, the isolation of suburban communities and the opening of large one-stop supermarkets, paid employment has largely changed the practices of food shopping. When combined with the growth of car ownership and associated growth in household mobility, people nowadays are able to travel a bit further for their regular food shopping, although many of them still shop locally. Increasingly, the car is seen as a 'shopping basket on wheels'. In this circumstance, spatial factors like distances and directions are less relevant because the time spent on travelling is not much different, especially when such bulky shopping trips are made less frequently (probably once a week or every fortnight). This is most common in the outer suburbs of London, though there are no shortage of similar examples in Inner London.

For example, Mrs Gerton, who works part-time in Harrow town centre, prefers to shop in a slightly smaller supermarket, Waitrose in Wealdstone, rather than in a large supermarket, Tesco, just a few minutes' walk away. Likewise, Mrs Bevan likes to shop in Wapping (Safeway) rather than in the local Tesco. In Wapping, by contrast, Mrs Wallis tends to shop in Surrey Quay (Tesco) rather than in the local Safeway. For them, the distances, directions and locations of the shops are not a problem at all; these shops are all *local* shops in terms of time–space proximity. For example, Mr Sand, a full-time househusband, prefers to shop in different supermarkets to check the prices.

Given that the time available for food shopping has decreased considerably due to the fragmentation of everyday life, it is very common that food shopping trips need to be combined with other trips. For example, Mr and Mrs Slade tend to do their major weekly shopping in Colindale on Saturday morning because their daughter has a playgroup there every Saturday morning. So, while their daughter is playing with other children, they get their weekly shopping done.

As many interviewees noted, they do not have a fixed pattern of shopping trips but depend on the possibility of combining shopping with other trips. For example, Mr and Mrs Wexford tend to share the responsibilities of food shopping and they normally shop together. As Mrs Wexford explains:

We have no fixed patterns or routes . . . as long as it is convenient. It really depends, depending on which way we are going. If we go to the West End then we will shop in Safeway in Barbican; if we go to Kent [to see relatives] then we will shop in Asda in the Isle of Dogs. Otherwise, we will do it in Safeway here in Wapping.

Households appear to be increasingly flexible and varied in their shopping practices as a means by which to coordinate different daily movements more efficiently in space and time. It is increasingly common to find employees using lunchtime breaks to shop near their workplace, or they shop on the way home in shops at transit locations or at places near home, or in the evenings or at weekends when they are combined with other trips. In short, the shopping practices of ordinary households are shifting away from a pattern of either regular major shopping trips or more frequent visits to local shops to a more diverse, more dispersed pattern of shopping trips involving a variety of shops at different locations and times.

Home, work, shopping and transport: the problem of meaningful integration

Changes in shopping practices are not isolated phenomena, but rather closely related to other aspects of daily life. The shift in food shopping trips from small independent shops in town to one-stop large supermarket chains at free-standing locations, for example, and from weekdays to evenings and weekends must be understood with reference to associated lifestyle changes in employment, housing and transport.

Given London's scale, neither unchecked decentralisation nor return to more concentrated retail development can meet the diverse needs of contemporary shopping practices. Shopping trips are increasingly governed by flexibility and variety with respect to location, size, type, opening hours, transport links and so on. The problem is how to ensure this potential for choice (when, where and how to shop) benefits households equally according to geography, gender and resource base.

Not only does the spatial structure of current retail development matter, but also the temporal rearrangement. For example, the most significant change in practice in recent years has been the introduction of late-night opening and Sunday trading. This contributes towards the vitality of the '24-hour city' and might be argued to attract people to live and work in the inner parts of London. By virtue of the fact that women traditionally take a larger share of shopping responsibilities, and they also comprise a larger proportion of the workforce in retailing, such changes in shopping practices also have profound housing and employment implications.

The extension of opening hours to late evenings and Sundays contributes to fresh demand for greater numbers of part-time (predominantly female) workers. This then has associated implications for housing and labour market structures. Women account for more than 60 per cent of the total national workforce in British retailing, half of which are part-time (Kirby 1993: 196). Accordingly, the trend towards longer opening hours in retailing has the potential to reduce the time available for workers to be with their families (Gottlieb *et al.* 1998: 31). In short, changing patterns of shopping bring about new coordination problems, particularly with respect to home, work, family, transport and child care. Therefore, it is necessary to consider the whole range of household life in a holistic time–space context, linking everyday household life and the overall urban structures.

Conclusions

This chapter has shown that the life of cities can be viewed through two distinct lenses – that of the institution (labour market, housing market, etc.) and that of the household, that is, the broader structural properties of cities and the mundane and 'trivial' daily practices that shape, adapt to, and reshape those structural properties. Moreover, different structural properties relate to each other via the institutional webs across larger and smaller time–space zones. This institutional interconnection is important not only for the production of urban structures, but also for the very reproduction of urban spirit: the major forum of social interaction. The life of cities and the livelihood of urban inhabitants are an integrated whole. Planners and policy-makers focus all their attention on the formal, structural aspects of cities; little attention is paid to, or information collected on, the users.

In the case of London, however, evidence from both primary and secondary data suggests that Londoners are living increasingly fragmented lives which are caused by, and reinforce, the structural disparities between employment, housing, retailing and transport structures in both space and time. In a sense, these time–space mismatches are closely related to a broader concern of sustainable development in a number of ways, notably via the generation of more and longer journeys made by cars and other motorised vehicles. By virtue of the mutually impinging character of household life and the interrelated nature of institutional structures in the cities, it could be argued that individual measures implemented at the margin of sectoral boundaries are unable to address the necessary relations between sectoral goals. Rather, an integrated, holistic approach is required to overcome the barriers of institutional coordination. Here, integration is not simply an interchange between bus and rail, but 'joined-up thinking' about provision and use.

The UK government is proposing an integrated strategy of coordinating transport and land-use planning in the existing urban boundaries for the pursuit of a sustainable future (this will be discussed in Chapter 6). However, it could be argued that the human scale of coordinating household dynamics is an indispensable part of the broader time–space coordination between institutional structures. This microhousehold aspect is, in effect, both the building block and the ultimate goal for an effective channelling of institutional structures. This is especially important for those disadvantaged groups that are more likely to live against the grain of cities. Arguably, greater understanding of the interconnection between household dynamics and urban structures is the key to exploring the very nature of modern cities and, accordingly, greater coordination of fragmented household lives and mismatched urban structures is the key to managing sustainable urban development.

Notes

1. For statistical purposes, Greenwich is sometimes classified as an outer borough and Haringey and Newham are classified as inner boroughs.

2. The information required for the analysis of London's institutional structures is mainly drawn from official statistics, especially the decennial Census of Population (the latest census was conducted in 1991) and the biennial or triennial Census of Employment (a sample census). Other official statistics from departmental offices, such as the former Department of Transport and the Department of Employment, and the London Research Centre (LRC) and London Planning Advisory Committee (LPAC) are also the major sources of data collection.

3. Between 1983 and 1993, for example, only an annual average of 2.1 dwellings were completed per 1,000 population in London, compared to 3.6 dwellings in England as a whole (LRC 1995: 3). The number of new dwellings built in London has declined from almost 34,000 in 1971 to just under 15,000 in 1994 (*ibid.*: 87). Accordingly, it seems unlikely that London's housing structure will change dramatically in the next few years.

4. The average floorspace of a Metro Tesco store is smaller and, most importantly, there is no car park.

Chapter 4

Opening up the household

Introduction

This chapter explores the 'intimate' web of connections binding actors and environments together in three overlapping spheres of restructuring – those operating in labour markets, housing markets and gender relations. These spheres of restructuring share the household as their key point of intersection. We argue that it is through the household 'lens' that individuals form preferences, make decisions and cope with change concerning employment, residential location, spatial mobility, transport mode and range, child care, parenting and relative attachment to community and place. These dimensions of 'everyday life' for 'ordinary' working family households represent the practical, spatially and temporally constrained mediation of home, work and community. By exploring these hidden performances of everyday social reproduction, this chapter examines the way routine arrangements shape the fabric of urban structure and institutional environment.

We pay particular attention here to the 'inner workings' of the household, a theme set in motion by urban sociologists Lydia Morris (1990) and Sandra Wallman (1984) with reference to the restructuring of unemployment and poverty. Here, this theme is advanced by casting light on the reflexive constitution of gender relations and divisions of labour within social and kin networks, within the locale, through the evolution[1] and negotiation of particular household strategies. Qualitative examination of household-coping strategies, together with the contexts within which these strategies are reproduced, are suggestive of important dimensions of connectivity between home, work and family life (Jarvis 1999: 225). We understand the household to be usefully conceived, following the principles of structuration, as a duality of structure. This way of viewing individual and household behaviour, as part of an organic household–locale 'nexus', is pursued with reference to the practical connections between household decision-making and particular urban contexts. For instance, by tracing the routes to individual employment decisions, working hours and journey to work, group housing consumption and gender divisions of paid and unpaid work, it is possible to observe the multiple, crosscutting elements making up the crucible of intrahousehold negotiation.

Connections between housing consumption, labour market production and household reproduction are particularly apparent for the rising number of dual-earner households populating advanced industrial economies. In London in the 1990s, 57 per cent of working family households (couples with one or more dependent child) had two earners. Half of these were households supporting two full-time earners and almost half again comprised two full-time earners employed in professional or managerial 'careers' (Jarvis 1997). This trend is by no means unique to London or to 'world cities' generally. Indeed, higher rates of dual-earner and dual-career households are recorded in US cities, particularly in accessible outer urban neighbourhoods (England 1993). Yet this remains a potent force for change. Increasing international competition in labour market production coupled with growing house price divergence between urban growth centres and less buoyant local economies are contributing to stronger polarisation between 'work-rich' and 'work-poor' households and neighbourhoods. London represents the most extreme manifestation of this trend in the UK at the dawn of the twenty-first century.

The growing concentration of paid work within particular household cohorts also contributes to renewed rigidity in local housing and labour markets. This points to a critical paradox, overlooked in debates concerning economic globalisation and the emergence of world cities. In essence, the global push towards greater labour market flexibility has implications for reduced labour mobility. This is due to the spatial constraints imposed on multi-earner households in which two sources of employment are supported from a single location (Abbercrombie and Urry 1983; Snaith 1990, Jarvis 1997: 524). Some individuals can break out of this spatial constraint and take up employment beyond the reach of a socially or economically fixed residential location but the private costs are high and the numbers small. Green *et al.* (1999), for instance, identify a discrete rise in 'dual-location' households, in which one household member commutes weekly over distances that preclude daily contact with spouse, dependants and permanent home. Long-distance living offers those who can afford two homes a way to accommodate temporary or more permanent employment relocation as an alternative to the often cited 'trailing spouse' syndrome (Bonney and Love 1991; Bruegel 1996). Extreme house price differentials and high transaction costs contribute to the emergence of such 'alternative' home–work–family spatial arrangements. Tensions associated with spouse employment and practical or emotional attachment to a particular neighbourhood also play their part. Overall, however, dual-earner households, especially those representing low- to middle-income earners, are recognised to be less geographically mobile than single-earner households (Bielby and Bielby 1992; Green 1995; Jarvis 1999).

Before considering the most obvious implications of this tension between global market restructuring and the internal workings of the household, we need to generate a better understanding of the relationship between household employment structure and spatial and temporal arrangements underpinning local housing and labour market interaction. It is necessary first to demonstrate how the 'household lens' effectively amplifies the social implications and local contexts of global market restructuring. In this chapter we develop the notion of the household lens in three stages or arguments. First, we peer into the crucible of household decision-making to identify some of the key connections between housing, employment and household gender relations. Second, we develop a rationale for viewing the household as a duality of structure. Third, we apply this understanding of reflexive and reconstitutive household–locale relations to practical issues confronting the household everyday. We conclude by highlighting key implications for current housing, transport, planning and labour market debates, viewing global market restructuring through the household lens. This starts to address some of the paradoxes evident in contemporary urban structures and political economies.

Revisiting ordinary urban households

The research introduced in this and the following chapter isolates a subpopulation which parallels the 'ordinary urban families' featured in Elizabeth Bott's (1957) seminal study. This subpopulation comprises white 'nuclear' family households, defined as a different sex couple living together with one or more dependent children. It focuses on nuclear families with one or more partners in paid employment, resident as owner-occupiers living in comparable metropolitan urban areas. While this subpopulation represents only 20 per cent of all households in the UK at the present time, it is important to acknowledge that women's (and men's) everyday experiences of home–work–family spatial arrangements vary by age, race, ethnicity, class, sexual orientation and situation within particular household structures (Katz and Monk 1993: 12).

Because experiences vary so pronouncedly, it is important to control for these variables rather than attempt to explore the inner workings of a general population of households. A strong case is made for observing the behaviour of a specific subpopulation of households. It provides the means by which to clearly differentiate observations of gender and labour restructuring *within* households from observations of demographic and employment differences *between* households (Jarvis 1997: 528). It is well documented, for instance, that the birth of a first child largely corresponds with a temporary or permanent change in female labour force participation (Henwood *et al.* 1987) and, hence, with a restructuring of gender roles and household divisions of labour. Less well documented are interactions between the number and ages of children and parental working times and commitments at subsequent stages in child development (Katz and Monk 1993: 18). For some working families it is the birth of a second child, when the first is attending pre-school, when child-care costs and schedules preclude the continuation of both parents in full-time employment. For others it is when children are in their formative school years, not yet able to travel independently but attending multiple after-school activities, that coordination of home and work fall apart. Working parents can find that their older (articulate and persuasive) children expect *more* of their time, to watch them play at sport or to take them to dance class. Often the timing of these activities continues to conflict with employment demands, especially where employment entails business travel and antisocial hours. By selecting a specific population of households sharing similar life-course characteristics (raising a family) the dominant effect of the life-course on household decisions and decision-making is greatly reduced.

There are further arguments for focusing attention on a household form which, while frequently favoured by politicians and moralists alike, remains a minority structure, one which many commentators view as anachronistic. The 'traditional' family of the 1950s Hollywood sitcom is permanently on the wane. Nevertheless, the majority of individuals pass through a nuclear household structure at some point in the course of their lives (Somerville 1994). The fact that this passage is not always enduring owes much to the tensions experienced within households, where gender roles and divisions of labour are continually negotiated and renegotiated.

Trends of household formation, dissolution and reconstitution reflect changes in lifestyles, life expectancy and relations of gender and generation. We see the result in delayed marriage, postponed childbearing, non-childbearing, divorce, lone-parent households and renewed pressures for more flexible housing and labour market solutions. It can be argued that by developing our awareness of the very real and practical tensions operating within a specific 'nuclear family' stage, we gain greater understanding of household structure transformation. When the household is recognised as a site of conflict (of gender roles, labour divisions, family goals and over the allocation of resources), it can be appreciated that one of the outcomes of intrahousehold conflict is family household dissolution. In pursuing this course, it must be recognised that household structures, divisions of labour and gender roles all 'flow' across time and space. We need to view change as the product of a combined evolution of daily moments, milestone transitions and life-course processes. In this chapter, therefore, we introduce household 'biographies' to the discussion. These go some way towards capturing the iteration between taken-for-granted daily routines, exogenous events, 'unconscious' personal developments and processes of change across time and space.

Here and in the following chapter, narratives are drawn from intensive household and neighbourhood research[2] to illustrate the ways in which households cope with change in housing, employment and gender relations. The interviews generate the personal and employment biographies of each individual along with 'couple' decisions concerning housing and family formation. They focus on the processes of negotiation which go to make up paid and unpaid resource contributions as well as the manner in which this resource exchange shapes relative bargaining power in household decisions. Population and neighbourhood samples are selected from London and Manchester to achieve close comparison. The subpopulation of working family households is further selected and categorised according to household employment structure. For analytic ease, three 'idealised' employment structures are introduced representing 'traditional', 'flexible' and 'dual' households. These are a simplified form of the almost limitless array of possible working-time combinations in households with two adult partners. 'Traditional' male breadwinner households consist of a male in full-time employment with an economically inactive female. 'Flexible' households consist of a male in full-time employment with a female in part-time employment. 'Dual' (earner/career) households constitute those in which both partners are in full-time employment. Where possible, the occupational status of full-time work is identified so that dual-earner households describe those in which both partners are employed full-time in any one of the social economic groups (SEGs): IIIN skilled non-manual; IIIM skilled manual; IV partly skilled; V skilled and members of the armed forces. Dual-career households, then, represent the higher-profile minority of couples both occupied full-time in well-paid professional or managerial employment (SEGs I and II).[3] The imposition of this template usefully exposes those household narratives that break the bonds of typological conformity.

For some time now, research and debate has been moving away from a neo-classic economic treatment of the household as an unproblematic 'individual' consumer (Manser and Brown 1980; Hindess 1988; Hodgson 1988). Increasing sensitivity to household gender relations can be attributed, in part, to visible changes in the profile and geography of employment in the UK and other advanced economies. Since the 1960s, increasing numbers of women have been entering, remaining in, or re-entering the labour force after marriage and the birth of a first or subsequent child. At the same time, there has been a profound restructuring in the nature and extent of employment available to, and taken up by, both men and women. Deregulation and the proliferation of temporary and insecure employment characterise visible dimensions of this process. Individual and household 'strategies' are varied in response to the process of change. A typical response is for individuals and households to increase their exposure to paid work. The 'family wage' and 'male bread-winner' models are largely consigned to history. Indeed, many would argue that it is the absence of a 'living wage' which generates precarious social and economic relations for a significant proportion of urban households. Consequently, if key urban issues are to be meaningfully addressed, theory and analysis of household behaviour must abandon notions of 'homo-economicus' and the household-as-unit. It is to be welcomed that new research has begun to take the household as a key unit of analysis. Yet it can be argued that this focus does not go far enough. Many essential aspects of urban living remain hidden. We need to view the implications of global economic restructuring through the household to better understand individual behaviour and house-hold arrangements worked out every day as local solutions to change.

Peering inside the 'black box'

Despite increasing emphasis on the household rather than the individual as the key unit of analysis, existing explanations of household behaviour typically assume that preferences underpinning consumption and behaviour are 're-vealed' in action and decisions are consensual. These explanations keep closed the 'black box' of intrahousehold reproduction, sacrificing this hidden domain to measurable observations of revealed preference and aggregate behaviour. Preferences and decisions are viewed as finite events rather than as culled fragments of formation and making – processes involving significant inter-personal negotiation. In practice, many preferences and decisions do not reach the arena of performance because they are modified, postponed or thwarted through interpersonal negotiation and environmental influence. Peering into the crucible of decision-making, it is clear that households are not consensual despite normalised public displays of unity.

The household is a significant site of potential conflict and certainly one of perpetual, often breathtaking, feats of balancing. The biographies provide a tapestry of discourse and practice, capturing the way individuals and house-hold groups construct their identities and the boundaries of their milieu in 'cooperation and conflict' with others (Schutz 1967; Harre 1979; Bourdieu 1987; Eade 1997: 30). This enduring process of negotiation is clearly evident in discussions that go on in households over time. Conflict might result in an eventual 'winner' or, through continued non-action, a zero sum gain (for a sustained attempt to apply game theory to intrahousehold decision-making, see Carling 1991, 1992; Jordon *et al.* 1994). Equally, it can result in the emergence of new, hybrid, compromise preferences that evolve as emergent properties of repeated and rehearsed practice and discourse.

Many decisions are never made and thus do not result in revealed action. This is a very important point to make with respect to housing and labour market interaction. Decisions that are thwarted because of tensions *within* households (competing careers, locally specific child-care provision and deep neighbourhood attachments) effectively represent a hidden driver underpinning change (and resistance to change) in cities. This is made apparent in the household biographies introduced in this and the following chapter. The narratives provide evidence of the 'messy' nature of preference formation and decision-making. Husbands and wives do not talk about how they live out their daily lives with reference to discrete decisions about where and how to live or about which of them could or would earn how much doing what occupation, how they would spend their money, if and when they would start a family and how they would manage their lives around the constraints imposed by child-rearing. Instead they provide narratives of 'doing' and 'being' in a series of overlapping, sometimes inconsistent stories of past, present and future. This is demonstrated in the narratives through implicit reference to gender roles, in explanations that current roles and divisions came about 'naturally':

> I don't think we really discussed it, we just knew each other well enough to know what it would be like [raising a family] – Mr Loader ('traditional').

> It was something that really just happened, wasn't it? – Mr Livingstone ('flexible').

> He's very flexible, he's not a sort of typical male in as much as the wife is going to stay home and look after the children. I don't think he'd have ever married me if he was because he certainly wasn't going to get that – Mrs Mallory ('dual').

This unconscious, as well as more material, negotiation of roles, divisions, preferences and decisions is particularly well expressed in the following two narratives. These effectively describe key household decisions 'in the making'.

Mr and Mrs Linklater: 'I tend to be the stodgy one'

Sheila (35), a full-time general practitioner and mother of two pre-school children, describes a discussion she has been having with her husband for some months without resolution. Geoff (36), currently employed part-time in a hospital consultant post, is frustrated by the lack of progress in his career and the difficulties he is having finding a suitable full-time post within easy commuting distance – one which does not disrupt his wife's career or his children's education and which contributes more significantly towards the couple's heavy mortgage commitments. He pronounces quite firmly his desire to 'get on in life'. Sheila recognises that her husband would be 'quite happy to go off and work in Canada or Australia', an option which she does not wish to contemplate. She goes on to talk about the differences in their propensity to pursue change and take on the 'great upheaval' of relocation or a house move. She admits that she is 'less adventurous' than her husband and observes that 'there's not exactly conflict, but I think I tend to wrinkle my nose and pooh-pooh any wild, mad ideas, so I think I tend to be the stodgy one'.

This discussion has been rehearsed for nearly two years with no 'revealed' out-come. The issue is not simply one of competing careers or even asymmetric attachment to a particular house or locale – it is an ongoing definition of family household well-being.

Compare this with the following story describing one couple's social (and spatial) construction of preferences over time. As with the previous couple, Mr and Mrs Langham live in east London and similarly represent a 'flexible' household employment structure comprising one and a half breadwinners.

Mr and Mrs Langham: 'Circumstances have changed now'

Barry (38) and Joy (36) were not long out of secondary/junior high school when they first met. They were working at the same east London tool shop. Barry worked as an apprentice in the tool room and Joy worked in the office – two separate worlds. Both worked full-time for eight years before starting a family, contributing jointly to the mortgage on their first home. Joy was used to earning more than her husband and was at first reluctant to give up her new job, located in a prestigious building in the City financial district. Consequently, Joy's mother, who lived nearby, stepped in to provide full-time care for the couple's first child.

This arrangement continued for nearly three years. After the birth of her second child, however, Joy found travelling 45 minutes into the centre of London too tiring, and her mother was having difficulties looking after two children in addition to her own home. Barry suggested Joy take up a local secretarial position, which he had learned about through his personal network of friends. This job offered part-time (school) hours, considered by both to be 'very handy'.

The shift from two full-time to one and a half incomes coincided with Barry's promotion within an international manufacturing firm and it effectively curbed rising tensions surrounding child care and domestic work. Mrs Langham explains, however, that working part-time and earning 'peanuts' compared to her earlier salary has not diminished her voice in key household decisions. She suggests, for instance, that she can veto an overseas move which would benefit her husband's career. She can do this because she speaks both for herself and her children and as the preserver of household and extended-family stability. She claims:

> I already turned him down once, to go to Belgium; I didn't want to go. We had the chance of going there for about two years, but I didn't really want to because of the kids, settled in their school and because I had a job and I wouldn't have a job when I got back and I'd miss my family.

Mrs Langham goes on to explain, however, that this was not a fixed veto, that circumstances are changing and her husband has persuaded her that a temporary overseas move might bring benefits that she had not thought of previously:

> Circumstances have changed now, where my family are OK, doing there own thing if they have the chance, and I've got problems with next door [an escalating neighbour dispute], so, I said if there's a chance of us going abroad, which did come up just before Christmas, to go to Germany for two years, I would go because he convinced me that the schooling was better out there and we'd be able

to save up a lot of money and I could always get a job while the children were at school.

The reality of an overseas move has yet to materialise, but discussion has evolved over a period of years and this continues to shape the way external opportunities and constraints are received.

Not only do these narratives reveal the 'messy' nature of preference formation and decision-making, but they also start to expose the intimate connections between overlapping issues within spheres of home, work and gender relations. This connection is never more apparent than in the relationship between housing costs, household employment structure and arrangements for social reproduction (most notably child-care provision). Take, for example, the case of Mr and Mrs Leicester.

Mr and Mrs Leicester: 'But life was so different then'

Both Rob and Jo grew up within a mile or so of their present home. They met in their late teens when Jo was at technical college taking a course in book-keeping. Rob left school at sixteen and undertook a series of 'odd jobs' before getting more permanent work as a minicab driver. A child came along earlier than planned, when Jo was still at college, and, as a consequence, finding 'decent' affordable housing became a key issue. Despite working nights and clocking up significant overtime (working up to 60 hours per week), Rob could not earn enough on his own to buy a house in the area in which they had grown up. The following is an extract from their joint description of this time.

Mr Leicester: I think at that time [sixteen years ago, when first married] the men went out and, but things have changed now, what with the mortgage and the situation with money.

Mrs Leicester: For housing.

Mr Leicester: People need two incomes and I think the women can command more money at the moment, quite honestly, don't you?

Mrs Leicester: Yes. [. . .] I don't think you've consciously made the adjustment [. . .] but, there's no housing anywhere, and you can't. I remember when I went to the housing advice centre; we were living in a two-bedroom house and we'd just had [third child] and we were renting, and I went up to see them about the chances of a council house and, literally, it was twenty-odd years, and you're lucky, unless you're actually homeless they're not interested, so unless we both worked–

Mr Leicester: –and buy somewhere, we had no chance.

Mrs Leicester: We had no chance of having room enough for our family. I think that's true. [. . .] One of us couldn't get by with working; it's both of us. We need both of us to work. We couldn't get by with one earner now. It makes you wonder how we got by in the beginning. But life was so different then.

It was in order to meet high London housing costs that Rob and Jo constructed a dual-earner family life. They circumvented the prohibitive costs of private child care by taking opposite shifts in their respective jobs. For twelve years Rob worked nights while Jo worked days. They left each other little notes explaining how the kids had faired and what still needed to be done. Consequently, Rob and Jo describe their current division of labour in a manner which cannot be materially or conceptually divorced from spatially and temporally constituted housing and labour market events.

The biographies provide numerous examples of this interdependency concerning housing, employment and household structure. Couples differentially perceive the 'need' to sustain one, one and a half or two full-time earners and the position taken in this regard feeds into housing choice (location and position on the 'ladder'). Equally, housing market events engage with employment behaviour and household gender divisions of labour. What are 'dismissed' in the (auto)biographies as routinised (taken for granted) practices amount to intersubjectively constructed representations of past, present and future 'lived experience' (Maffesoli 1996; Durrschmidt 1997). For example, individual preferences are frequently masked by capitulation to external events, limited options or 'given' circumstances:

> It was never considered an option that I'd stay at home – Mrs Mistry ('dual').

> We always knew we were going to have children. I don't know whether we'd discussed whether I'd give up work. At that time, when we were both full-time, I was earning more than [husband] and I don't really think we could've just said I'd stop working. I don't think that was an option – Mrs Lively ('flexible').

Furthermore, it was previously observed that the binding together of housing, employment and gender relations demonstrates the co-evolution of these three spheres. This suggests the 'situatedness' of household practices, strategies and discourse within locally embedded social and kin networks. The case of Mr and Mrs Land illustrates this process.

Mr and Mrs Land: 'I don't think I would have chosen to go back'

Peter (36) and Sally (36) met each other at school. They both grew up in central east London with parents who had bought their own homes and always worked locally. They bought a house and moved in together a year before they married, paying for the mortgage jointly from their two full-time wages. Peter worked on an early morning milk delivery round and Sally, who took basic nursing training after secondary school (to State Enrolled Nurse level), worked in a local hospital. They talked about starting a family as soon as possible but Peter was concerned about his low wages and the fact that his job was leading nowhere. He continued his milk round by day and took vocational engineering classes at night.

With improved qualifications Peter was able to apply for his first 'salaried' employment as a trainee engineer. By this stage Sally was pregnant with twins.

The best-paying job prospect involved relocation fifty miles to the Essex coast but this move was short-lived. Sally felt cut off from her family at a time when she wanted practical support in caring for newborn twins, one of whom had been born with significant health problems.

They thought through the implications of returning to the city. At first they were shocked to discover how much house prices had escalated in the two years they had been away. Peter took a better-paying job in Central London and for some months he commuted for more than an hour by train to work each day. But this did not help the couple save for a move back to east London and, with Peter gone such long hours, Sally had even less help with the twins.

Finally, in order to secure affordable accommodation, they bought a house in need of extensive modernisation. The cost of undertaking this modernisation, even as an exercise in self-provisioning, required greater income than Mr Land's new employment could provide. Consequently, it was this move back to east London which triggered Sally's return to paid employment. At this point in the narrative, Sally explains:

> Once I had the twins I didn't really think about work, full stop; I don't think I had time to [. . .] When we moved back into London, because we needed a bigger mortgage, that's why I went back to work, basically, to pay the difference [. . .] but, basically, I don't think I would have chosen to go back. In fact I was quite depressed when I first went back.

A gradual transformation of gender roles and divisions of labour took shape over a period of five years. This was such that when a third child arrived, Mrs Land returned to full-time employment after only six weeks of maternity leave. While Sally at first returned to work three days a week, she subsequently pursued further vocational training (advancing to State Registered Nurse status). This increased commitment was partly in response to restructuring within this occupation and the reorganisation of shifts at the hospital in which Sally worked. In order to work at her preferred location (close to home) and on her preferred shifts (regular days), she recognised the need to improve her occupational status. However, through this process she increased the number of hours she committed to paid work. Mrs Land continues:

> I only switched to full-time when I did my course in 1993. Before that I was doing three days a week and then the two days I had off I was looking after my sister's boy, from the age of five months to fifteen months, and I found I got more stressed on my days off looking after him than when I was going to work. So, plus I was wanting to do this course, so I felt I needed to up my hours.

While she increased her hours, she still claimed that, for her, 'the children come first' and that this marked the difference between hers and Peter's working lives. Mr and Mrs Land go on to describe this continuing negotiation of home, work and family life:

> Mr Land: Well, from a personal point of view, we've not been having arguments, but I've been having pops at her about her promotion, you know, them promising her this–

Mrs Land: —but the thing is, I can find a [better] job but it means . . . at the moment it takes me ten minutes to get to work, and it's convenient.

Mr Land: I know, but I keep saying you gotta draw the line.

Mrs Land: But it's not important, is it?

Mr Land: It is important, in principle.

Mrs Land: Yes, but the children have to come first.

Several connections and interdependencies emerge from this narrative. Connections are made apparent between the evolution of household employment structure, the situatedness of household practices within social and kin networks, and the negotiation of strategies of mobility or, as in this case, the consolidation of a particular house and locale as being a permanent place of residence (Jarvis 1999: 235). The spatial proximity and intensity of social and kin networks provides one element in the 'glue' that binds housing, employment and gender relations together. By locating closer to her parents, Mrs Land gained access to essential help with child care. She also helped out two days a week looking after her sister's pre-school-age child. It was through the process of moving back to the more expensive London housing market, close to family, that Sally returned to paid employment. Her sister was in a similar position, contributing a second income towards household housing costs. Consequently, Mrs Land's return to paid work was at one and the same time facilitated and necessitated by returning to London.

What also emerges from Peter and Sally's story is a clear sense that 'moral economic' issues enter the crucible of decision-making through the arrangement and ordering of daily living. Though working full-time away from home, Sally sets an upper limit to the distance she will tolerate between work and home and the ability to reach her children should they need her at short notice. Equally, by consolidating a permanent place of residence close to extensive social and kin networks, Sally feels able to distribute direct responsibility for the welfare of the children more widely within the locale.

The 'push' and 'pull' factors influencing residential location and mobility are multiple and difficult to unravel. Individual motives are bound up with group negotiated strategies. It is unrealistic, therefore, to model gender roles, divisions of labour and household structure according to either structurally determined explanations of 'economic efficiency', 'time availability' and normative socialisation (Parsons and Bales 1956; Becker 1981; Hartsock 1983; these approaches reviewed in Hiller 1984) or voluntarist approaches explaining behaviour in terms of relative human capital investment. Neither is it adequate to ascribe primacy to cultures of patriarchy such that women are 'always' seen to be exploited (Oakley 1974; Hartman 1981). The narratives suggest that gender and power relations, articulated through preference formation and decision-making, are not defined by 'nature' or 'nurture' or by 'structure' or 'agency' in any predetermined manner (Davis *et al.* 1991; Watts 1991). Rather, individual and group behaviour is shaped through interaction. This suggests that, in order to capture the interaction of household gender relations, the household needs to be viewed as a duality of structure.

Viewing the household as a duality of structure ───────────

From the narratives we see clearly that if the secret lives of households are to be revealed, it is necessary to abandon the notion of a unitary body or agent. The notion of duality implies an institution that is at one and the same time a structure, one shaped by the activities and motives of constituent agents, themselves organised by this organic structure. This captures the multiple forms and plurality of contexts of household gender and power relations. It largely builds upon Arendt's (1970) understanding of cooperative conflict, as currently applied to larger political economic entities. In this way the household is recognised as an institution of both objective form (money from wages, capital, property, skills, state-legitimised authority over dependents) and subjective relations of thought and perception (shared customs, conventions, morals, beliefs, norms). Individual agents occupy positions within the household institution interdependently with the group structure itself. Each household member separately participates in spheres of activity inside and outside the nominal boundaries of the household action space or 'home sphere'. These activities and social interactions, constituted through the discourses and practices of daily life, are the emergent 'reality' of the institution. The separate participation of individuals in external activities and social relations, such as paid employment or trade union activity, is not appropriately conceived as functioning within a 'separate' 'public sphere' because of the interdependency of the practices, discourses and power relations of individual agents and the household group.

Feminist theorists have long argued that the 'separate spheres' of work and home are artificially defined. In practice, patterns and inequalities of power and status in public life will filter through to the private sphere (and vice versa) (Edley and Wetherell 1995) through the ideological reconstitution of space. Saltzman Chafetz (1991) describes this transference as occurring through a boundary that is asymmetrically permeable. Whereas superior male power in the public sphere maintains a strong separation of work and home (through home-centred child-care policies and labour market assumptions of productive male workers being supported by 'invisible' partners providing reproductive labour 'off-stage') (Hochschild 1990; Massey 1995), females in paid employment experience a blurring of the boundaries between work and home in terms of both roles and expression (Pleck 1979; Wheelock 1990a; Saltzman Chafetz 1991). Equally, it can be argued that men, who are working long hours in jobs demanding adaptive flexibility, also frequently experience a duality of these spheres where male preference for a greater parenting role is hindered by the structures of workplace regimes (Cohen 1993).

It is thus recognised that the practices, habits and routine arrangements at the heart of daily living are both reflexive and recursive within and between spheres of activity and networks of social interaction and knowledge. The implication is that households are situated within a shared system of dispositions, practices and learning. This 'situatedness' percolates through individual and group-lived experience alike. Rather than having a unifying effect,

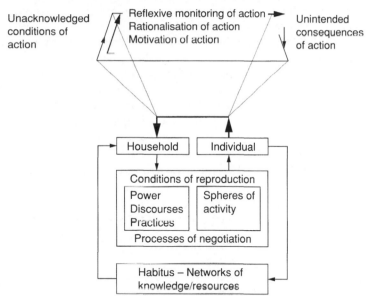

Figure 4.1 The relational link between household members and the system of the household

however, individual agents remain 'free' to introduce innovation to the group at the same time that the group is able to influence the receptiveness of that individual to innovation. Attention is given to sets of dispositions and practices which are 'enduring rather than eternal' (Bourdieu and Wacquant 1992: 133). Dispositions are 'practical in character' and express 'the capability to "go on" within the routines of social life' (Giddens 1984: 4). While dispositions (tendencies towards particular routines, practices and ways to 'go on') may be associated with particular household structures, the former is not sufficiently realised by the latter. Consequently, it is through the unfolding of the biographical narratives that it is possible to operationalise these key elements of the duality of the structure of the household.

Figure 4.1 represents the interdependence of the individual household member, as a reflexive agent, and the social reproduction of institutionalised household practices. The illustration integrates a well-established conceptualisation of agent reflexivity (Giddens 1984: 5) together with a new conceptualisation of household–locale association. Agent reflexivity is represented as the magnification, in a planometric projection, of the relational link between household members and the system of the household. While the household, as a structure, is unreflexive in this relationship, agents reflect upon the household system in a way that is differentiable from self-monitoring. Moreover, interdependence is not brought about through reflexivity alone. The household is more than the sum of its (reflexive) individual members. The emergent 'reality' of the household is constituted through the mediation of multiple overlapping institutions (including class and gender). These institutions, overlapping within and between the membership of the household, mediate the distribution and interplay of power (material, moral, emotional) and resources (money, property, knowledge) through daily household practices.

We can see then that the household orders individual reproduction (consumption, production and reproduction) at the same time that practices and resources, drawn upon in reproducing social action, provide the means of reproduction for the system of the household as a whole (Giddens 1984: 19). More importantly, this process is constituted through relations of interdependency whereby the spatial and temporal situatedness of households serves to reconstitute that space (home, place and locality) and time (embodiment and movement). Household action (interdependent reproduction of structure and agents) entails the continuous flow of reflexive monitoring, by individual agents, of self, other and the system of the household as a whole across time and space. As with individual action, much occurs which is unintended or unacknowledged. Consequently, the rationalisation of household action may be represented ad hoc or post hoc by individual household members in contradictory ways.

The practice of transformation

The narratives provide evidence that households function as a duality and that this understanding provides a constructive analytical framework for household and neighbourhood research. One example is suggested in the way discourse and practices associated with home, work and family life are exchanged within and between households in a particular locale. This exchange of norms, habits, 'novel' ideas and cautionary tales charts a process of 'evolutionary learning'.[4] Another way of looking at this is to see households as using particular tactics, or strategies, to 'get by'. Particular rules and resources are available to individual household members (as 'holders' of different forms of knowledge) and household institutions (as mediators of networks of knowledge reproduction). This process of mediation ensures that networks not only provide access to a variety of forms of knowledge, but they also exclude or 'screen out' alternative sources of information. A transparent example of this process is provided in the following extract from Mr and Mrs Lampton's ('traditional') story. This couple describe the way their approach to parenting has been influenced by Mrs Lampton attending a series of evening classes with friends:

> Mrs Lampton: The first year I attended the college I did child psychology, and I think that actually had a huge influence.
> Mr Lampton: On both of us.
> Mrs Lampton: Initially on me, but, ultimately, as parents really, plus there were friends who were in a similar sort of field, got me interested, so I did this child psychology course and just got involved in it more and more and it just made me think so much and I'd come home and we'd discuss it.

This narrative describes the transmission of 'new knowledge' (ideas relating to child psychology) between Mrs Lampton's social networks (friends interested in this subject), Mrs Lampton's discussions at home with her husband, the couple's combined experiences of parenting and Mrs Lampton's contributions to the evening class and social networks. While this is quite an explicit example of the transmission of practices and knowledge concerning parenting, these relations of interdependency between the situated individual and the situated household are implicitly revealed by couples throughout the narratives.

A further example is suggested by the informal exchange of ideas, practical help and material resources between households. Mrs Lively describes this with respect to the unpaid help she receives from her parents, who regularly look after her children, and the reciprocal support she offers with respect to her parents' failing health. She observes that

> A lot of problems can be solved with a close-knit family. My mum's my [helper], dad even; my dad goes down the school and does hobby hour with the kids. They're the backbone, aren't they [to husband]? They're just there all the time, and it works both ways. I don't think it's all one way.

She goes on to describe the amount of help she and her husband have received in the course of a complete renovation of their former local authority home. Mr Lively is anxious to point out that 'it works both ways' with the home improvements too, because, as a mechanic, he is always willing to donate his time and skills as a return favour. This reciprocal relationship is quite parochial (though not always geographically 'local'). It ensures that certain values, skills and habits are captured and reproduced within the network:

> Mrs Lively: We've not paid for anybody to come in and do work. My uncle came up from Bournemouth for three consecutive weekends plumbing for us and my cousin is an electrician and rewired for many weekends. We supplied the beer and the eats, didn't we, and lots of people, lots of friends have helped.
>
> Mr Lively: It's mainly favours, because if any of them damage their cars they phone me to fix them up, so it's handy really, no money changes hands, but then I know that if they damage their car, they'll phone me up and I'll sort the car out for them. It's good; it's like a big extended family.
>
> Mrs Lively: That's something you didn't understand, coming from the other side of London.
>
> Mr Lively: I didn't at first, but, I did find it a bit hard to understand, but once everyone gets involved, it's quite good. So, but my father-in-law, he's a site agent, so he has the knowledge. Whereas, at the other house I learnt a lot from my father-in-law, I was able to do a fair bit on my own [this time].

The understanding that individual and household behaviour is influenced by the existence of social and kin networks is not new to urban sociology (see Bott 1957 for one of the earliest discussions). Indeed, it is common to identify and 'measure' such networks in terms of qualitative interdependency and reciprocal exchange (Gouldner 1960), proximity and intensity (Hirsch 1981; Belle 1982) or the 'strength' of the ties between members of a network (Granovetter 1973, 1985; Granovetter and Swedberg 1992).

What is generated in this discussion, however, and not fully developed in existing research, is the suggestion of a key link between the processes of transformation within and between situated households and wider systems of change shaping the fabric and political economy of cities. In effect, networks of knowledge, learning and experience can be seen to function on multiple and interconnecting levels. These represent the parochial scale of the babysitting circle; the 'local' level of housing and employment markets; the cultural terrain of shared history or religion; the national surveillance of government regulation; the 'global' impact of international divisions of labour and capital. Scope for transformation exists within and between each level. This permeability is such that global effects clearly percolate through national political economies to translate into local responses and vice versa. One way of revealing this relationship going on 'behind the scenes', as it were, is to observe the essence of everyday life, the minutiae propelling 'ordinary urban households' one day to the next. First, it is necessary to understand the relationship between space, time and the everyday coordination of individual and household action.

Space, time and everyday coordination

Imagine for a moment that it were possible for each of us to trace the path of our day, making this visible to a cartographer in outer space. This exercise would be something in the manner of the young Tom Sawyer trailing his kite-line when making his way through the passages of a cave to ensure that he can safely find his way home (Mark Twain 1876). This would be one way of mapping the spatial arrangement of an individual's day. It would reveal the relationship between the activities of each household member and others in their immediate social network. Though such an exercise is highly improbable, it is possible to visualise the resultant image (it might be compared in scale, for instance, to Christo's[5] 'surrounded island' sculpture which involved 6.5 million square feet of pink fabric – used to wrap around eleven islands off the Florida coast). From outer space, some kite-line maps would be very tightly woven to reflect many crossings and connections within a local geographic area. Others would be dispersed over a large area with few connections or intervals within the space of a single day. Some might represent a very singular existence; others the responsibility for a great number of tasks and other people's lives. It might also be possible to establish differences on the basis of gender, class, ethnicity and the extent to which individual and group social reproduction is sustained by informal, state or private-market welfare provision.

To understand the full significance of the points of connection and crossing in the kite-line maps, it is necessary to recognise that each and every day (and part of each day) is not only spatially arranged but also closely ordered by time. Further, the temporal ordering of each day typically follows a 'critical path' whereby events have to occur in specific sequence or are bound by a fixed timetable (pre-school opening hours, school times, work hours, medical appointments, shop opening hours). As we noted in Chapter 2, these largely describe the principles of time-geography first expounded by Torsten Hägerstrand (see page 38). Arguably, the narratives point to key relationships between household decisions and 'strategies' concerning housing, transport, employment and child care and processes of spatial arrangement and temporal ordering underpinning the social reproduction of everyday life. When we start to view individual time-geographies through the household lens, we start to see a new set of relations emerging, those associated with the coordination of individual and group social reproduction. In Chapter 5 we generate a better understanding of spatial arrangements and temporal orderings for each of the 'idealised' household structures identified earlier. At this point it is sufficient to point out the lessons to be learned for urban social policy and planning from a better understanding of the principles of time geography orchestrating home–work relations every day (for further details of this analytic approach see Hägerstrand 1976; Rose 1993: 17–40).

It is easy to see how the combined spatial arrangements and temporal orderings of urban populations contribute to key issues in contemporary urban social and economic planning. One example is suggested by the crosscutting trends underpinning the mode and route by which children get to and from school. We introduced this debate in Chapter 1. It is instuctive to revisit the key issues to view these more particularly through the household lens. In an interview in June 1999 John Prescott (Minister responsible for the Department of Environment, Transport and the Regions) commented on the serious environmental damage caused by the 'school run' and increasing use of the private car for short journeys. Such trips, which might previously have been conducted on foot, by bicycle or bus, contribute indirectly to global warming and atmospheric pollution as well as traffic congestion in cities. If we view this issue through the household lens, a number of paradoxes emerge. This is not simply a trade-off between private car use and environmental policy. The suggestion is that parents are being overprotective (Moorhead 1999). No mention is made of the very tangible time constraints entailed in getting a child to and from school (or more than one child to and from different schools, in different locations and with different start and finish times), having to fit essential chaperoning around the working schedules of one or more parents working long or irregular hours. It does not acknowledge that, with increasing numbers of women in paid employment propping up stagnant family wages and meeting international demands for flexible labour, time for the school run is necessarily squeezed with the result that greater recourse is made to the private car. Increasing car use is also associated with the dissolution of traditional school 'catchments'. It also reflects more generally the privatisation of family welfare under a neo-liberal regime whereby the private car has replaced the school bus. Furthermore, parents are encouraged to shop around for 'better' schools beyond their immediate neighbourhood. Consequently, fewer children attend the 'local' school. Children of the same age, in the same street, may not attend the same school. In this way, privatised transport solutions have also become individualised, where once they might have generated informal collaboration between neighbours. None of this is to say what ought to underpin policies concerning transport mode and use, education choice, family welfare and norms concerning working parents and child care. The point is that the spatial arrangements and temporal orderings underpinning the school run, as with all moments of everyday social reproduction, bind spheres of activity together in such a way as to make what might superficially appear to be a 'transport issue' one which is profoundly enmeshed in home, work and community relations.

Spatial arrangement and temporal ordering

Clearly, the household lens is not simply made up of social relations but also spatial relations. Interaction is spatially articulated in terms of ideologically sanctioned 'action spaces' (kitchens for cooking, bathrooms for bathing, bedrooms for sleeping, etc.). Household gender relations may be articulated differently according to a range of spatial performance levels. For instance, time spent travelling (as a function of distance) between paid-work and domestic-work 'spheres' is limited by the demands of the dual roles and shifting between task-specific spaces. It is also widely recognised that there remains an unequal spatial gender division of labour (Duncan 1991; Massey 1994). Women's increased participation in paid labour has not disrupted the perceived separateness of the home as a reproductive space. The narratives demonstrate the impact of practical constraint on the constituents of household structure and vice versa. Spouses exercise power in negotiating the relative 'permeability' of home–work boundaries (Hartsock 1983) at the same time that local labour markets generate very real tensions, in the form of working hours, working times and mobility imperatives, for household gender relations.

Elsewhere, it is noted that local concentrations of male-type occupations are associated with practical constraints to female spouse employment, for example through increased domestic workload and scheduling difficulties (Massey 1995, p. 191). Further manifestations of the spatiality of power are evident in spatially embedded cultures of 'gender at work' (McDowell 1997: 137). In these ways, the regional concentration of particular occupational profiles and local labour market structures both contribute to and are shaped by local cultures, norms and expectations surrounding 'male' and 'female' divisions of work (Duncan 1991; Hanson and Pratt 1995). Furthermore, it can be argued that the boundedness of public and private spheres prevents a greater gender equalisation in household divisions of labour because this persistent ideological separation of productive and reproductive functions imposes a 'double-shift' on women combining work in the two spheres (Geerken and Gove 1983; Hood 1983; Finch 1983; Hochschild 1990). This boundedness also typically limits the choice of paid work of working mothers to 'local jobs' (Hanson and Pratt 1991). Yet little has been said about the scope of diversity in the way gender and power relations are negotiated in practical contexts of spatial and temporal constraint.

Household gender relations are not only negotiated across space, but also time. Pragmatic temporal and spatial coordination typically provides the impetus to, and sustains the boundaries of, household decision-making. By way of example, the narratives demonstrate that power is not only mediated in relation to household structure (who does what in terms of paid work) but also who does what where within given scheduling arrangements and working-time conditions. Consider for a moment some of the practical experiences underpinning the 'school run' discussion above. Anyone who has tried to chivvy a young child into putting on their shoes and donning hat and coat with any sense of urgency when the clock is ticking and a fixed appointment awaits will know that time is experienced quite differently by children and adults. It is also experienced differently by the same actors according to relative circumstances of formality, autonomy, enthusiasm or apprehension. By speeding up and intensifying paid work for many working parents, global market restructuring is effectively speeding up the transition from home to work. This might not translate to a shorter commute (in terms of time or distance) and thus greater spatial concentration of home place and work space. For some this is clearly 'what gives' in the squeeze, but for others spatial rationalisation is not an option or it is rejected because it conflicts with other priorities. This speeding up is more likely to translate to intensification of this transition. Commuters are often seen on the train conducting their first hour's work on a lap-top computer and mobile phone before reaching the office in the morning, and a final hour's work on the way home. In the same way, it is evident that many working parents drive their children to school to speed up and intensify their transition to work. Where previously 'child-time' might have directed the walk to school (stopping to prod at a dead insect, jumping the cracks in the pavement) 'firm-time' dictates that harassed working parents spend 'car-time' ensuring that their children know the schedule of the day, which parent will pick them up and what clothing to have packed for which after-school activity. While there is not scope to pursue the qualitative dimensions of time in this chapter (captured in fashionable reference to 'quality time'), it is important to recognise that spatial arrangement and temporal ordering reflect non-material (moral and emotional) norms as well as material (economic) imperatives (for a comprehensive discussion of the 'time-bind' confronting working parents see Hochschild 1997, 1990 and for 'time as a commodity' see Schor 1992).

Arguably, a full analysis of household decision-making and everyday practice must take account of the spatiality and temporality of gender and power. One way of viewing the influence of temporal ordering is to study the timetables associated with particular household structures (such as dual-earner households), identifying the timing and coordination of each household member's paid and unpaid work activities. This approach was used by Pratt (1996a) to illustrate the integrated nature of people's everyday lives. It is difficult to identify from this approach alone the conditions and motives underpinning 'revealed' spatial arrangements and temporal ordering. As we shall see in the following chapter, what is considered to be sustainable in terms of home–work–family spatial arrangement and what is acceptable as a timetable for these 'separate spheres' varies significantly between individuals and household structures.

Conclusions

This chapter has focused our attention on the relations within and between households and the 'secret' performances binding individuals, households and localities together in a practical sense. Household 'strategies' and ways of coping with change in these overlapping spheres are understood to reflect the spatial arrangement and temporal ordering of home, work and everyday family life. The biographical approach that we used makes it possible to 'make public' these hidden performances within households, as well as those linking households in a particular locale. These private and informal performances (unpaid child care, the reciprocal exchange of skills and resources associated with home improvements, income generating activities undertaken in the 'shadow' economy) too often remain hidden in existing urban studies.

We have emphasised the way households construct and reproduce the boundaries of their lived experience. This is described in terms of a shared system of dispositions, practices and learning – representing the spatial and temporal situation of households within an evolving 'modus operandi' (Krais 1993). It is through reflection on past events and engagement with overlapping institutions that the household reconstitutes existing norms, habits, discourse and practice. The argument is that by viewing the household as a duality of structure it is possible to undertake household and neighbourhood research which then illuminates the principles of structuration in action. Narratives effectively describe the 'flow' or 'becoming' (Pred 1985: 338) of household structure and the way this in turn shapes the environment of this reconstitution. Narratives drawn from a specific subpopulation of working family households have been used in this chapter to illustrate two key themes. First, we identified the household as a significant site for research. We introduced the frame of household gender relations as a lens through which to view wider processes of restructuring in terms of both theoretical explanation and method of analysis. Second, we explored the 'inner workings' of the household by illuminating the *formation* of preferences and processes of *decision-making* and negotiation underpinning decisions to act. We identified multiple 'push' and 'pull' factors shaping housing choice, number of household earners and working hours, child-care provision, journey to work, residential mobility and gender roles and divisions.

The second of these themes is further developed in the following chapter when the question is asked why similar households respond differently to exogenous trends of social and economic change. The following chapter also explores in more detail the tensions between global economic restructuring and local practices of arrangement and ordering. Here it is sufficient to conclude, from our journey from theory into practice, that viewing restructuring through the household lens clearly demonstrates how routines, spatial arrangements, schedules and practices shape and are shaped by urban structure and institutional environment. The situatedness and embodiment of individual household members within overlapping local social networks implies that decisions and actions that are typically orchestrated within the household sphere are constantly being disrupted by positions of status and influence negotiated beyond these intimate relations. This embeddedness of intimate relations in a wider locale, often representing a political economic environment over which individuals have little control, suggests that space and place can provide solutions to the coordination of everyday life, as well as rigidities to the way households might respond to wider economic restructuring. This suggests that one unintended outcome of the globalisation of modes of production, and proliferation of various forms of labour market flexibility, is the 'localisation' (and privatisation) of strategies of social reproduction and welfare (Beck and Beck-Gernsheim 1995: 35).

At the beginning of the twenty-first century, urban societies in advanced economies are confronted with many difficult choices. The treadmill of expectation concerning continued economic growth and the rigours of international competition often run counter to the preservation of geographically unique and finite resources. In the UK, a transparent example is provided in the current debate over the allocation of land for new housing. Consultation following the Green Paper *Household growth: where shall we live?* published in November 1996, has raised local concerns about how the dwellings necessary to house the additional 4.4 million households, projected to form between 1991 and 2016, are to be allocated across the country (Jarvis and Russell 1998: 6). Concern over the quantity, location and distribution of land for new housing has led to widespread belief in the 'compact city' and of urban containment as a primary planning objective. New households are not forming (or migrating) evenly across the country and demand for new housing is most concentrated in the south-east, particularly at the edges of conurbations, and in the Shires, where pressures for containment are greatest. London is quite unique in experiencing significant pressures for housing in the urban centre, as witnessed by press amazement at the prices paid for loft apartments in EC1 (Clerkenwell, the late 1990s 'place to be'). Elsewhere, the poor environmental quality of urban centres is exacerbated by continued out-migration (Champion and Ford 2000). The result is a highly uneven patchwork of both regional and highly localised housing markets ('hot spots' and 'black holes'). Allied to this is the clear mismatch between income and house price (or rent) distribution and, consequently, home–work spatial relations.

As we noted in Chapter 1, many workers in low-wage essential services (nurses, teachers, transport staff) are being squeezed out of accessible residential areas. For dual-earner households this potentially further strains already complex spatial arrangements. A real paradox is that London's bull economy is providing employment for more minimum-wage workers (cleaners, caterers, security guards), as a multiplier of expansion higher up the income ladder, and this growth is driving up house prices and rents beyond the reach of low-wage workers. Furthermore, policies of urban containment, established on the grounds of environmental concern, can contribute to high and rising house prices in core areas (where housing supply is essentially fixed by containment) and thus indirectly to further out-migration (leap-frogging the green belt) and to extended commuting. It is a feature of housing development in the UK that large green field housing developments at the edge of conurbations, detached from compact public transport networks, are more likely to yield affordable housing than centrally located brown field sites. There are implications here for the exacerbation of patterns of income and social polarisation, the fragmentation of home–work spatial relations as well as for social and environmental sustainability over the long-run.

The implication is that tensions exist for urban planners and policy-makers, which are in large part mediated at the household level. Yet it continues to be the case that planning for housing fails to take account of the way households resolve these tensions through the spatial arrangements and temporal ordering they negotiate. Urban planning and social policy typically views the household as a unit, without recognition of the duality of this structure and the local embeddedness of household practice. By failing to view urban issues in the round, through the household lens, planning and social policy initiatives aimed at simply consolidating home and work are likely to fail.

Notes

1. The term 'evolution' is used as a metaphor for organic transformation rather than to fix our understanding of social reproduction within a Darwinian or Spencerian frame of 'natural selection'.
2. This project combines original quantitative and qualitative research in a two-tiered methodology. Qualitative research centred on 30 in-depth interviews conducted with 'nuclear family' households in two comparable neighbourhoods (identified as around 200 homes): one in Barking, east London, and one in Prestwich, north Manchester. In each neighbourhood, five interviews were conducted with each of the three idealised household employment structure types: 'traditional' (single male breadwinner); 'flexible' (one and a half earners); and 'dual' (two full-time earners).
3. Clearly, there also exist asymmetric 'cross-class' households in which husband or wife hold a higher-level occupation than their spouse (McRae 1986). However, these are a minority and represent less than 8 per cent of the subpopulation of working family households under observation here. While cross-class households were not specifically selected out of the sample of households interviewed, the research picks up only same-class couples and discussion largely assumes this condition. Once again, by controlling for the variable class it is possible to throw a more consistent spotlight on to issues of change in gender roles and relations.
4. To some extent this notion of 'evolutionary learning' corresponds with an understanding of institutional reproduction. The emphasis on 'evolution' conveys the paradoxical tension which operates between stability of habit and change through interaction. There is in this discussion a strong parallel with Bhaskar's (1989) 'Transformational Model of Social Activity' (TMSA) in which society is viewed as 'an ensemble of structures, practices and conventions that individuals reproduce and transform' (p. 76) (see also discussion in Pratt 1995). Reproduction (though not a process of direct replication) represents a continuity, while transformation offers disruption and mutation of either a moment in, or trajectory of, continuity. These ideas work harmoniously with the principles of structuration and offer scope for further development within household research, both theory and practice.

5. Christo is a Bulgarian-born 'installation' artist and sculptor living and working in California, USA. He is famous for undertaking projects of great enormity as a means of 'humbling' space and making elements of the natural environment accessible to the human scale. Projects include extending a 'curtain' between two mountains in Colorado, 'wrapping' islands and filling the Arizona desert with giant umbrellas, to name but a few.

Chapter 5

The strategies of social reproduction

Introduction

For many years, urban planners, residential developers, politicians and sociologists alike have desired access to what they regard as the secret world of preference and behaviour in housing and labour markets. Why (and how) is it that households sharing common social, economic and demographic characteristics behave so differently in equivalent market conditions? Without exploring the inner workings of the household it is impossible to address this question or to explain variation in the spatial arrangement and temporal ordering of home–work–family relations. Taking detailed observations of household gender divisions of labour as the 'key' to achieving access, and recognising the household as a site of negotiation, this chapter explores axes of differentiation in urban housing and labour market behaviour. Interviews with working family households illustrate key points of tension emerging from these overlapping spheres of decision-making.

This project anticipates that the sustainability of particular home–work–family arrangements hinges on the combined nature and extent of work undertaken by individual household members. In the first instance, then, spatial arrangements and temporal orderings differ according to household employment structure – whether, for instance, working family households comprise one, one and a half or two full-time earners over the long-run. Here, what is considered 'sustainable' in terms of residential mobility, journey to work, child-care provision and family welfare both reflects and shapes household gender divisions of labour. This process is clearly conditioned by local opportunities for male and female labour as well as by cultures 'naturalising' particular ways of 'doing' gender and caring for young children.

Often omitted from this 'balancing' equation are the very practical constraints facing households with more than one earner. In this case the location and timing of work undertaken by one partner impinges on the employment opportunities of the other. Obvious examples are those jobs which impose shifting working hours, long hours, irregular hours on call-out, overtime on evenings and weekends and periods of work away from home overseas. It is a neglected fact that these 'flexible' employment practices offer mixed blessings for household employment coordination (Hanson and Pratt 1995; Massey 1995). Consequently, relative flexibility to the needs of spouse employment and to child care need to be viewed through the lens of particular household gender relations. Shift work, for instance, tends to increase male spouse availability for child care (time spent as the primary carer in a dovetailed work schedule) but it also heavily circumscribes female spouse employment options and prospects. In the 'Ottawa' shift system, for example, which is commonly employed in the police force, shifts rotate between 'earlies', 'lates' and 'nights' on a five-week cycle. This fluctuation of shifts prevents a spouse taking up regular (office hours) employment in circumstances where paid child care is ruled out. Equally, residential mobility is far easier for two earners both working in relatively ubiquitous occupations. For those employed in a specialist field or working for a unique employer, household relocation is less likely to benefit two partners equally. This might lead to the 'sacrifice' of one career or inertia for both.

A 'strategy'[1] of permanent household location emerges in the case of Mr and Mrs Linklater, introduced in the previous chapter. Mr Linklater describes the specific 'pull' of London as a location suitable for cultivating two careers (in medicine):

> It was always on the understanding that with [both] of us trying to find, particularly training posts, you tend to have to wander round different hospitals, that London has enough pull that [we could] both get jobs, while smaller cities don't particularly have all the jobs available.

The decision to first locate and then to assume permanence in a metropolitan employment 'hub' is particularly characteristic of couples with two partners employed full-time in professional or managerial positions. Nevertheless, this case provides salutary evidence of the perverse constitution of labour market opportunities. Mr Linklater is currently unable to find full-time work in his specialist field. If he is to advance his career he effectively needs to 'wander round different hospitals' over a far wider geographic area, a prospect which is greatly at odds with the strategy of coordination originally negotiated in this two-earner household. More generally, new varieties of 'feminised' and 'casualised' working hours and working times are increasingly replacing 'traditional' (male) full-time, secure and continuous employment contracts. Recognising this trend, we wish to argue here that the *nature* (quality, terms and conditions) of paid work is as relevant as the participation (hours worked) of female spouse employment. Only by viewing the *combined interaction* of working hours, times, opportunities, conditions and spatial arrangements *within* particular household structures is it possible to unravel the contradictions underpinning housing and labour market interaction in cities.

   This chapter is organised in two parts. The first part introduces a thematic analysis of a complete set of interviews and household housing and employment histories assembled for 30 working family households interviewed in London (Barking) and Manchester (Prestwich). This highlights key patterns of association linking the 'idealised' household employment structures introduced in the previous chapter to particular strategies of behaviour. The question is asked whether practices differ between household structures in a sense beyond that demonstrated across society as a whole. While any attempt to categorise households by 'type' necessarily reflects a simplification of reality, it is important not to view preferences and decisions as entirely idiosyncratic. The following discussion strikes a concerted balance, therefore, between a constructive search for shared experience on the one hand and sensitivity to diversity on the other. Revealed axes of differentiation then form the agenda for a closer reading of the household biographies in the second part. Central to this discussion are alternative interpretations of 'risk-sensitivity', 'rootedness' and 'flexibility'. Constructions of 'risk sensitivity' hinge on perceptions of personal insecurity and risk and those of 'rootedness' the propensity for individuals and households to coordinate home, work and family life through locally embedded strategies. Questions of 'flexibility' pertain to the practical constraints confronting households combining working hours and times. Household arrangements are examined with particular reference to households with more than one earner. This highlights the practical negotiations underpinning urban housing and labour market interaction in cities experiencing both rising numbers of dual-earner households and the proliferation of new 'flexible' forms of working. The chapter concludes by highlighting some of the wider implications of a localisation of welfare in households coordinating multiple sources of income.

Two urban neighbourhoods

Intensive neighbourhood research offers a means by which to apply 'real life' insights from everyday household practices to explanations of housing and labour market behaviour. It recognises that a comprehensive picture of household decision processes requires an understanding of non-events as well as events, of that which is taken for granted as well as that which is discussed and contested. It is only through intensive primary qualitative research that these more elusive facets of household behaviour are made apparent. Wallman's (1984) ethnographic study of eight London households remains the closest proximation to this mode of investigation (see also Tivers 1985; Morris 1989,

1990; Snaith 1990). While much contemporary sociological research focuses attention on household divisions and negotiation, these accounts remain largely aspatial (see, in particular, Hochschild 1997; Nelson and Smith 1999). Here, a 'bio(geo)graphical' approach makes it possible to advance existing theory and debate concerning the effects of economic restructuring on household practices. We trace the ordering and interplay of joint and individual housing and employment events through the medium of narrative. The aim is to expose both cross-sectional and longitudinal interactions and thus increase our understanding of gender and generation combined.

Included within this project are 30 in-depth interviews conducted with an equivalent subpopulation and an identical range of 'idealised' household types living in one of two residential locations. This subpopulation of 'ordinary' urban households represents different-sex partners living together with one or more dependent children. At least one partner is in paid employment living as an owner-occupier in older, second-hand housing (inter-war, 1918–39 three-bedroom semi-detached). Households with a single full-time male breadwinner are classified as 'traditional' structures. Those with one and a half earners are labelled 'flexible', named after the flexible labour market practice of female part-time employment, which is the identifying feature of this household type. Households with both partners in full-time employment are referred to as the 'dual-earning' type. Where possible, this type is further differentiated by occupational status. While interviewee occupations largely fall within the low–middle income status group, a number of 'dual-career' households are separately identified. Here, the conditions of full-time professional or managerial employment confer opportunities and constraints that differ from those in which a second full-time earner effectively props up a low-paid 'earner'.

One of the neighbourhood study areas featured is situated in Barking, east London, and the other in Prestwich, north Manchester. These outer urban neighbourhoods offer contrast between different regional urban cultures at the same time that they share in common key local environmental conditions. Each neighbourhood provides access to rapid light-rail transport (London Underground and Manchester Metro) and good access by road, rail and bus to the city centre. Both offer similar quality and type of urban structure, housing stock and local amenities. Figure 5.1 illustrates the location and plan of each neighbourhood.

Not only do these neighbourhoods share similar physical appearance in terms of housing type, street layout and recreation facilities, but they also demonstrate equivalent residential 'status' and community profile. Despite the existence of older, cheaper housing, neither of these are local housing markets which could be described as undergoing any visible process of gentrification. These are long-established residential areas which, while offering a relatively central location and traditional urban feel, remain culturally detached from 'city' life. Both represent areas of low unemployment, high owner-occupation and a relatively homogeneous,[2] primarily white population with an above average proportion (35%) of 'nuclear family' households (ONS 1991) (see appendix for a more detailed discussion of the research method and design).

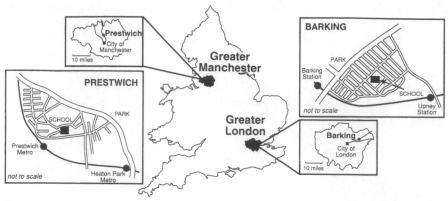

Figure 5.1   Location and plan of east London (Barking) and north Manchester (Prestwich) neighbourhood study areas

Comparative urban household research provides a powerful framework within which to observe sources of particularity and commonality in the way households coordinate home, work and family life. This approach highlights the contingency of local cultural contexts as a factor in explaining the adoption of different 'strategies' by households of the same employment structure. It is anticipated that individual cities (particularly two spanning the putative north–south 'divide') will reflect and in turn recreate differences in household housing and labour market behaviour. This is one of the questions addressed below. Yet this project exposes many common household practices. Shared experiences emerge in large part because of the high degree of specificity built into the research. If the project were to focus attention on the role of class or ethnicity, it is likely that further crosscutting associations would emerge and fewer shared experiences made visible overall. The decision to control for class and ethnicity here does not deny the importance of these variables. It is simply a means by which to identify sources of tension within particular household structures and local urban contexts.

Barking rests within the eastern fringes of London (in the London Borough of Barking and Dagenham). It offers one of the most pronounced examples of a local culture supporting the 'traditional' images of female 'homemaker' and male 'breadwinner'.[3] Duncan (1991) notes from his examination of secondary employment data for Britain that 'financial pressures (pushing women into the labour market) do not extend to the suburban east of Barking' (p. 103). This is understood to be due, in part, to the relative security of well-paid skilled manual jobs for men in the local vicinity, as well as to historically strong patriarchal influences of the male breadwinner as a 'macho' status symbol. Compared with all other districts in Outer London (a metropolitan area with above average rates of women in employment overall), Barking demonstrates the lowest rate of 'dual-earner' working families. Here, 43 per cent of nuclear family households have two earners compared with 53 per cent in Harrow (outer north-west London), the source of interview data introduced

Photograph 5.1 Typical inter-war housing in Barking, east London (Source: Helen Jarvis)

in Chapter 3. Not only does it appear that the local labour market reinforces 'traditional' gender divisions and roles in Barking, but it can also be argued that housing costs have contributed little to the 'need' for more than one bread-winner. Barking offers a market for older, second-hand single-family dwellings which are 'affordable' when compared with alternative provision elsewhere in London (see Photograph 5.1). In 1995 the price of an inter-war semi-detached house in Barking was 50 per cent cheaper than the equivalent prop-erty in Ealing, the same distance (west) from the city centre (Halifax Price Indices 1996). Notwithstanding the impression of gender role 'traditional-ism', which emerges as a 'snap-shot' from 1991 Census of Population data, working family households living in Barking today face continuing market restructuring. Households interviewed in 1996 spoke of the growing 'need' for more than one earner and difficulties associated with male unemployment, underemployment and reduced wage opportunities.

Prestwich lies on the northern outskirts of Manchester (in Bury Metropolitan Borough). This area, comprising the former cotton towns of the north-west region, has experienced a long tradition of women in paid employment (Massey 1994). While Manchester has fared poorly in the wake of economic restruc-turing, Prestwich has emerged from this process superficially unscathed. This is not a neighbourhood bearing the visible scars of deindustrialisation; empty houses and abandoned shops and businesses, as witnessed elsewhere in the region. It is a remarkably stable, prospering (though by no means affluent) residential area. The district of Bury records the highest rate of dual-earner nuclear family households in Greater Manchester, a metropolitan area representing above average rates of dual-earner households overall. In Prestwich, 62 per cent of nuclear family households ('couple' households living with one or more dependent children) represent a dual-earner structure (two full-time or one and a half full-time earners) compared with a rate of 43 per cent for the same subpopulation in Barking.

Table 5.1 highlights the profile of housing, employment and household gender relations in each of the two neighbourhoods. Overall, this data summary demonstrates the level of social, economic and demographic equivalence achieved through this particular neighbourhood selection. Rates of male employment (and unemployment) are broadly similar, as are rates of owner-

Table 5.1 Profile of housing, employment and gender relations in Barking and Prestwich

Housing Tenure %			
	Buying	Owner occupier	Private rent	Social rent
Barking	75	5	4	16
Prestwich	74	7	2	17

Housing Profile %			
	Detached	Semi	Terrace	Flat
Barking	10	34	40	16
Prestwich	16	44	36	4

Employment (Women in 'couple' households) %		
	Full-time	Part-time	Economically inactive
Barking	12	28	60
Prestwich	20	34	46

Household Structure (employment composition of 'nuclear family' households) %		
	Both employed	One employed	Neither employed
Barking	43	45	12
Prestwich	62	29	9

Socio-economic Groups (SEGs) Social class of economically active family head %					
	I	II	III(N)	III(M)	IV	V
Barking	2	18	18	36	18	8
Prestwich	7	32	15	29	13	4

Key to SEGs:
I = Professional; II = Managerial; III(N) = Skilled non-manual; III(M) = Skilled manual; IV = Semi-skilled; V = Unskilled. Members of the armed services have been excluded.
Source: ONS Crown Copyright, *1991 UK Census of Population* (100% file) and ONS (1994) *Key Statistics for Local Authorities*, London: HMSO.

occupation and general standards of living.[4] Notable contrast is achieved with respect to the propensity for women to work full-time, or at all, in Barking compared with Prestwich. By isolating the role of female paid employment, it is possible to observe whether the same 'idealised' household structures behave differently according to gendered cultures of production, consumption and reproduction.

Thirty working family households

Before exploring the particularity of the household biographies it is useful to identify those dimensions of everyday domestic practice that suggest typo-logical conformity. By constructing a template of this nature (not unlike a repertory grid) (Ryle 1975) it is possible to identify key relationships. This exercise provides a starting point for closer reading of the narratives. Figure 5.2 combines data for each household from individual employment histories, in-depth interviews with couples, 'diaries' recording who does what to sustain everyday life and key milestone events associated with spheres of home, work and family life. Home is defined in terms of the bundle of 'services' flowing from a fixed residential location. Within this theme the biographies reveal motives for housing choice and residential location as well as the propensity to view a particular house or neighbourhood as 'permanent'. This engages with individual and group willingness to contemplate future relocation or increased journey to work. Individual work histories reveal all aspects of paid employ-ment undertaken (hours worked, experience of unemployment and relative job security) together with unpaid domestic work, voluntary work and activities in the 'shadow' or 'informal' economies. Questions about family life reveal the proximity and intensity of local kin networks as well as provision made for child care and related aspects of everyday social reproduction. The suggestion is that each of the 'idealised' household structures typically employ strategies of behaviour that are characteristic of that type. From this template it is possible to develop a 'vignette' of each of the 'idealised' structures as follows.

Vestiges of the 'family wage' in 'traditional' male breadwinner households

'Traditional' households apparently experience the greatest security of male spouse employment as well as the most visible (and visibly gendered) separa-tion of paid and unpaid spheres of work. This household type is the least likely of the three to perceive a 'need' to prop up primary employment with additional paid work in the 'informal' economy. It is also the least likely to undertake a significant volume of 'self-provisioning' activities such as do-it-yourself home improvement and most likely to employ outside contractors to undertake home or car maintenance. Child-care provision is the exception. This and other aspects of care-giving remain exclusively maternal. Respons-ibility for accommodating everyday uncertainties and scheduling difficulties (sick dependants, school closures) falls to the female spouse. This is explained

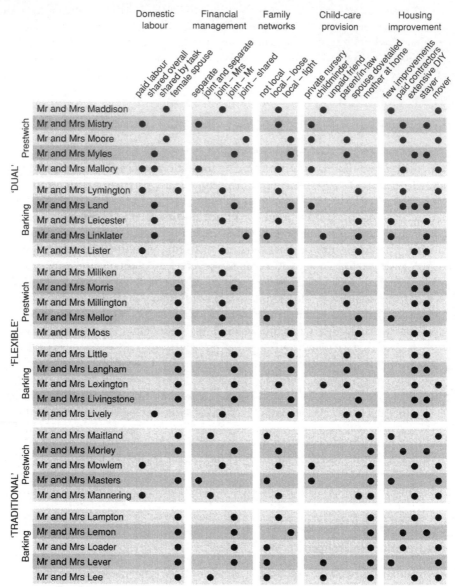

Figure 5.2 Schematic representation of domestic practices and strategies of coordination for thirty households, comparing three 'idealised' household employment structures

in part by the 'demand-led' working hours and timing of male spouse employment. Domestic labour typically remains a female preserve with limited contributions (by task type and time) from husbands. This pronounced division effectively preserves male spouse capacity for the long working hours required (and made possible) by sole breadwinner status. 'Traditional' households are more disposed to be geographically mobile than other types and express a general willingness to contemplate relocation if this were to benefit male breadwinner employment. Though a majority of these households remain in close proximity to one or more sets of parents, they do not rely heavily on local family and friends to sustain everyday domestic arrangements.

Extending resources in 'flexible' households

Greatest male spouse insecurity exists in 'flexible' households (a type representing 40 per cent of employed nuclear family households).[5] Couples in 'flexible' households in both survey areas expressed the 'need' for a second income either as an essential financial component or as an important 'safety net' to prop up insecure primary employment. They are more likely than other types to seek regular or periodic sources of additional income from informal economic activities ('moonlighting' for 'cash in the hand'). It is consistently in 'flexible' households that the majority of self-provisioning (home improvement) takes place. A common housing strategy is to purchase a 'run-down' property as a means of gaining access to an affordable family-sized house. This typically requires significant personal investment (own time, own labour, family assistance, materials and skills) and is associated with an expecta-tion of staying 'permanently' in the present house. 'Flexible' households are also the least likely to be residentially mobile. These households cultivate the strongest local social and kin networks and generally manage child-care provision by sequential spouse employment (dovetailing) or by drawing on unpaid help from a parent or other relative. In Barking, paid child care tends to be 'ruled out' because of a belief that children should be looked after only by a family member (an extension of 'traditional' mother-at-home values). In Prestwich it is more typical for paid child care to be 'not an option' because of its cost relative to income. Wives undertake the majority of day-to-day domestic provisioning such as laundry, cleaning, cooking and shopping, with husbands performing the role of 'weekend handyman' and of providing a 'taxi-service' for children's after-school activities.

Balancing gender divisions in 'dual' households

Households with two full-time earners typically employ some form of paid child care. They may live in close proximity to family members but are less likely to be in a position to draw on these networks to provide a regular source of unpaid child care. When 'dual' households are differentiated by occupa-tional status it is evident that 'dual-earner' households are frequently as reliant on unpaid child-care provision as 'flexible' households. This type, character-ised by low combined income and long combined working hours, generally relies on local informal assistance with child care, especially to make after-school and sick-child provision and to cover 'gaps' in overlap between spouse employment schedules. It is in professional and managerial 'dual-career' house-holds (comprising 10 per cent of employed nuclear family households) that child care is provided through the private market, outside the family home.

Furthermore, with two partners in full-time employment, spending extended periods of time away from the home, spheres of activity and social interactions are typically described as being separate and scattered. 'Dual-career' households demonstrate the least reliance on local networks and the greatest recourse to provisioning via commercial markets. Though less likely to contemplate relocation than the 'traditional' male breadwinner type, households with two full-time wage earners, particularly where these represent professional or managerial careers, are more disposed to be mobile (in terms of lengthy commuting and residential mobility) than 'flexible' households. In effect, the lack of 'rootedness', which is apparent in 'dual-career' biographies, both conditions and is conditioned by extended participation in activities associated with paid work away from the immediate locale (Dex and Rowthorn 1997).

Gendered urban cultures

Superficially, there is little variation between cities in the everyday practices of the 'idealised' household structures. For instance, 'flexible' households consistently demonstrate the greatest stability both in terms of 'type' transformation (unchanging structure over the life-course) and strong local attachment (inertia). Nevertheless, financial practices appear to differ more distinctly by region than they do by household employment structure. In Barking, the majority of households manage a joint bank account with husbands assuming financial control. This pattern of male control breaks down in 'dual-earner' households where wives are in many cases the higher earner and where their involvement in business and financial occupations explains the 'natural' (by skill specialisation) assumption of household budget control by these working wives. The Prestwich sample is characterised by a diversity of money management styles. Joint, separate and a combination of accounting systems are in use (Pahl 1988; Burgoyne 1990). Greatest financial control by wives occurs in 'flexible' households in Prestwich, where household budgets are considered tight and where four out of five wives assume financial control of a joint account.[6]

The key relationships

There are obvious limitations to any attempt to 'survey' or make a quasi-quantitative study of intrahousehold processes from biographical research. While there are apparent patterns of typicality for particular household types (and regional cultures), these are frequently disrupted. With the exception of 'flexible' households, there is little consistency to intrahousehold characteristics. For instance, households with two full-time wage earners (dual-earner/dual-career structures) share similar characteristics with regard to reduced mobility and more egalitarian domestic practices but when it comes to explaining processes of decision-making it is clear that typologies based on occupational status or cohort might provide a more consistent frame of interpretation. This is quite understandable given that any typology based on a cross-sectional 'snap-shot' of household divisions of labour will interrupt the 'flow' of household structure.

The meshing together of couple work histories changes through the family life-course such that a biographic approach captures a transformation through several 'types' longitudinally. We consequently need to pursue a closer reading of the biographies.

While intensive (and thus small-scale) research does not lend itself to any formal measurement of association, thematic analysis suggests that household strategies connecting home, work and family life hinge on three key relationships. These suggest routes by which to interrogate different and diverse preferences and behaviour in ostensibly similar households.

The first dimension pertains to the relative sensitivity households experience to risk and insecurity. Sensitivity might stem from actual experience of precarious employment (unemployment, reduced hours or wages), insecure housing (overcrowding, housing repossession or negative equity) or fragile transport or child-care arrangements. Equally, it can derive from the *perceived* worsening of a household's ability to 'go on' in a practical sense.

The second dimension can be described as the relative 'rootedness' of the household and the extent to which attachment to a particular house or locale is enduring. On one level, strong residential attachment might reflect the attractiveness of the local environment, for instance by providing a 'nice quiet place to raise a family' (Mr and Mrs Morris, 'flexible'). It might also stem from reluctance to embark on the 'upheaval' of a house move or relocation (Mr Lexington, 'flexible'; Mr Moss, 'flexible'; Mrs Millington, 'flexible'). Equally, deep attachments frequently stem from practical and emotional reliance on local social and kin networks. An enduring relationship exists between the scope and intensity of local networks and place-based practices and strategies. This was demonstrated in the previous chapter with reference to the Livelys. Mrs Lively's parents effectively bridge the 'gaps' in spouse employment schedules, providing child care and transport to and from school. This solution to coordinating two jobs is dependent on them remaining in close proximity to Mrs Lively's parents.

The third dimension relates to constructions of 'flexibility' and the practical options particular households face in coordinating the location and timing of home, work and family spheres of activity. It is suggested that each of these characteristics, such as proximity to local social and kin networks and the nature and extent of male and female spouse employment, function in combination to reinforce key axes of differentiation.

Coordinating housing and employment in two-earner households: 'risk sensitivity'

A paradox emerges from the thematic analysis above. It appears that the 'flexible' type (named after the defining flexible labour market practice of female part-time employment) is in fact the *least* flexible. It is the most consistent to 'read' but the least likely to adapt to the global economic imperatives of a mobile and responsive labour force. The relatively rigid position of this household type can be explained by its sensitivity to precarious employment.

Precariousness contributes to (and combines with) a strong practical local attachment. Significantly, today's 'flexible' labour market practices appear to operate hand in hand with personal insecurity and risk. Whereas at one time the state and the firm absorbed the impact of macro-economic boom-and-bust cycles, it is now the case that the household accommodates these vicissitudes. One manifestation of this is seen in the 'need' for working family households to support additional workers and consequently accommodate complex acts of balancing to coordinate two jobs from a single residential location together with child care and domestic social reproduction. These household practices and strategies of behaviour are sensitive to perceptions of long-run employment security as well as expected norms and the desire to maintain 'a decent standard of living'. The implications of this sensitivity are clearly illustrated in the case of Mr and Mrs Livingstone ('flexible').

Mr and Mrs Livingstone: 'If we were to have a decent standard of living, I had to go back to work'

Gary and Louise both grew up in the Barking area. They married 'very young' (19) and had their first child 'earlier than expected'. Louise did not intend to go back to work after starting a family. She believed that Gary could provide for her because he brought in a good wage as a self-employed bricklayer. She also felt that, because her own mother had been a full-time mum, she too should do the same. However, around the time the couple had their second child, Gary experienced financial difficulties. The construction industry was in deep recession and the most lucrative private-sector jobs had dried up. Concerned about the dearth of work, his effective drop in income and the precariousness of self-employment generally, Gary decided to switch to a 'regular job' – labouring in the public sector. This stabilised his hours and pay but left his income at about half what it was before. Louise recognised that if they relied solely on Gary's income the family would have to cut out holidays and work on the house. She consequently decided to go back to work part-time to compensate for this shortfall. She was reluctant for non-family to care for their young baby and pre-school child, so she took on 'bank' (agency) nursing working two nights a week when Gary was available to watch the children. This exhausting schedule continued for two years. Louise enjoyed 'having an income (of her own)' and decided to take on a more challenging 'career-type' job, working as a sales representative five days a week. This was much better paid and came with the benefit of a company car – giving Louise a great deal more mobility than previously, when Gary monopolised the family car during the week. When the children both started school, however, Louise found that a regular day job did not fit around school hours and she felt particularly under pressure during long school holidays. Gary was unable to help out now that he worked fixed hours for someone else. Moreover, he had started night classes, training to be a cab driver. Gary's strategy was to move into better paid 'freelance' work with prospects for more overtime. This strategy effectively limited Louise's employment options because she retained primary responsibility for the 'school run' and other family welfare matters. Louise left the better-paid job and switched instead to low-paid part-time clerical work in her children's school.

This story suggests that global market restructuring, particularly macro-economic vicissitudes, increasing competition and the deregulation of labour, contributes to new localised solutions to personal insecurity at the same time that they reinforce gender divisions and inequalities. This is further demonstrated in the case of Mr and Mrs Mellor, another 'flexible' household experiencing a strong economic 'need' to support two incomes.

Mr and Mrs Mellor: 'avoid putting all your eggs in one basket'

Dan comes from Hulme originally, Gloria from Prestwich. They chose to set up home in Prestwich because it was a 'nice area', rather than because of family attachments. Gloria was concerned that they had a solid and secure financial base before moving on – cautious but also ambitious. Two years before starting a family, Gloria left a full-time salaried clerical position to work part-time from home taking in payroll processing. She knew that the couple needed two incomes, primarily as a 'safety net' because Dan's job was insecure, but she also wanted greater flexibility to be home for a family. Working from home seemed to her the answer to having 'the best of both worlds'. Earlier in their relationship, Dan and Gloria worked for the same firm but Gloria left for another job when the workforce went out on indefinite strike. She wanted to ensure that the couple always had one secure wage, however low. Her philosophy was to spread the risk and 'avoid putting all your eggs in one basket'. Now, just as the couple are expecting their second child, Dan's job prospects appear to be at their most precarious. Dan's firm is moving to new premises forty miles away. Mr and Mrs Mellor have discussed the options and Dan is willing to accept redundancy rather than to move with the firm. Neither wish to relocate or have Dan take on a lengthy and costly commute. Gloria observes:

> It would mean longer hours and more time away from the home, which we're not happy about with a second child on the way, and I wouldn't be able to take a second job because he wouldn't be here to watch the children. So I said, well if that's the case, that you have to move, then they'll have to make you redundant because we're not prepared to hinge everything on that job [. . .] Although he's the major breadwinner, I don't want, it's like all your eggs in one basket, and that's what we'd become. He'd be on a higher salary, more time away from home, and I'd just become a wife at home and I don't want to do that.

As a consequence, Gloria has taken on additional work in the evenings at a local call centre. She feels the need to both increase and shift the focus of her work for two reasons. First, she sees a tailing off in her hours of self-employment due to processes of computerisation underway in the payroll sector and, second, because she wants more employment security, regular hours and a regular wage. Her strategy anticipates the need for Dan to retrain so that he can find local employment and thus enable the couple to fit their jobs sequentially around care for their children.

This case demonstrates clearly the way that gender relations frame what is possible and sustainable for households coordinating home, work and family life. This household interprets housing and labour market events with conscious reference to the 'need' to support one and a half incomes. This employment structure serves not only to militate against Dan's precarious employment and provide the household with a 'safety net' but also, more importantly, to reproduce negotiated gender relations. This reflects a particular ethos to the way home, work and family life are to be 'balanced'. Gloria concludes:

> I like to contribute more and have him share more of the home life. He would be, what's the word, what I say is we work to live not live to work, and not, if he took the move, you know, he'd just live to work and I'm not having that – life's too short.

The Listers and the Lymingtons

It is not only in 'flexible' households that the precariousness of casualised and deregulated employment is felt. The fact that the Listers now represent a 'dual' household is largely a function of the processes of local labour market restructuring. Mrs Lister has recently moved into a 'good job', full-time, but Mr Lister has experienced the steady erosion of his earning potential. The first time Mr Lister was made redundant from a seemingly secure position (unskilled employment 'on the buses'), Mrs Lister took up part-time employment in retail banking to provide the household with a financial 'safety net'. When Mr Lister was made redundant a second time, Mrs Lister increased her working hours, vocational training and work commitment. Facing redundancy for the third time, the couple joke about a reversal of roles. Mr Lister supported them both for the first half of their married life and now Mrs Lister has assumed this responsibility.

Similarly, when Mr Lymington's bakery business collapsed, Mrs Lymington returned to work, first working nights so she could still be available for their youngest pre-school-aged child, then increasing her hours to accommodate Mr Lymington's extended unemployment and need to retrain. They now both work full-time. They describe the personal changes involved in making the temporary transition from a 'flexible' to a 'female breadwinner' structure:

> Mr Lymington: It was very difficult, for me. The roles had changed, you know, I didn't earn from the April through to the October. I didn't actually earn any money because the business had gone under and the stamps weren't paid so I wasn't entitled to unemployment benefit.
>
> Mrs Lymington: It was just hard, very hard work, physically hard, and then coping with what [husband] was going through and plus there were a lot of financial pressures [. . .] It's very hard just to change your role all of a sudden, from me doing everything indoors, which I was, you know, two and a half days a week, of course I was

doing it all, and it was getting back to, kind of, leaving lists: can you do this, can you do that. It was very difficult, when he was feeling so vulnerable anyway.

Later on in the narrative, Mr Lymington explains that while he was unemployed he continued to look for alternative 'informal' sources of income to compensate in some way for his precarious position:

Mr Lymington: We had a Renault Espace that I used to do, you know, hotel runs and theatre runs with someone that I knew that had a business and stuff, all right, unfortunately we had to give the wagon back but it paid for something and that, so I wasn't (idle), you know, it saw us through and it's amazing what actually comes up.

Mrs Lymington: We've got a neighbour, very good friends, he travels a lot and his firm pays for a taxi to take him to the airport – black cab – and for a little while he got [husband] to do it for him, so they were very supportive.

Mr Lymington: Yes, that was, in terms of community support, stuff like that, that was good.

The most telling impression here is the sense that, under pressure, households depending on more than one income cling more tightly to a familiar locale. This response effectively ties them more permanently to a particular local market and feeds into the relative 'rootedness' of 'flexible' and low-paid 'dual-earner' households. Continued maintenance of a second income, even if low-paid, provides a hedge against uncertainty for these households. This strategy reflects the consolidation of a permanent place of residence (strong 'rootedness'), in close proximity to local social and kin networks, as a means of combating financial insecurity.

Rootedness

Evidence of differential dispositions of household 'rootedness' resemble, in part, existing behavioural explanations of residential mobility. Concepts of 'place utility' and 'context dependency' come readily to mind (Wolpert 1964; Lin-Yuan and Kosinski 1994). Similarly, the notion that individuals and households can be 'tied movers' or 'tied stayers' in relation to their attachment to the locale. This language has particular purchase for dual-earning couples (Bonney and Love 1991; Green 1995; Bruegel 1996). Here it is suggested that relative rootedness is better described in terms of a socially and spatially constituted 'network utility' rather than a strict attachment to 'a place'. This is not to deny that an element of 'rootedness' results from symbolic attachments, which are appropriately associated with place rather than locally embedded networks. Symbolic attachments frequently abound in the identification of individual tastes with familiar landscapes and regional cultural identity (accents, language) as well as personal nostalgia and family history. Nevertheless, it is rootedness within socially and spatially constituted systems of support that directly influence the household strategies unearthed here.

Hanson and Pratt (1991) note that residential location (and mobility) is largely unresponsive to employment change for most couples (p. 126). Evidence presented here for the late 1990s suggests that labour market restructuring and the intensification of paid work within a subpopulation of 'work-rich' households contributes to yet greater inertia and renewed rigidities in housing and labour markets. Location is often associated with existing family ties and an environment which is familiar. Yet it is interesting to observe the way professional and managerial couple households appear to differ in this respect. To militate against the potential conflict of competing career mobility, these couples frequently assume a permanent location in a 'hub' metropolitan area where, because of the scale of the potential employment market, future occupational mobility need not require physical relocation. The choice of this location is often quite independent of prior attachment. This strategy was introduced in Chapter 4 in the case of Mr and Mrs Linklater (both medical professionals). It also features in the case of Mr and Mrs Lexington (both local government officers). The Lexingtons naturalise their 'flexible' employment structure as being 'a happy medium' in balancing home and work. While London provides the largest potential market for both their respective careers, this location strategy effectively reinforces traditional gender divisions and extensive commuting costs.

Mr and Mrs Lexington: 'a happy medium'

Mark and Tina met at the local technical college, both training as local authority officers. Mark had lived all his life in the east London area. Tina had moved around the country as a child and only settled in Barking in recent years. Although they both work in the same field, Mark has moved extensively between local authorities within metropolitan London, consciously pursuing promotion, whereas Tina has remained with the same east London authority. Mark has consequently always undertaken lengthy journeys to work, up to an hour by car. Permanent relocation, to reduce Mark's commute, has never been a consideration because, in the long-run, it is London which is viewed as the labour market. Seen in this way, both Mark and Tina could potentially pursue work in any one of 33 boroughs. Tina admits, however, that she is committed to working locally. She views her shift from full-time to part-time work after starting a family as 'a happy medium'. She intends not to return to full-time work while her two children are of school age, though she has experienced discrimination in the conditions of her employment since taking a part-time job-share. She is reluctant to take on more hours or responsibility because this would entail evening meetings. Since Mark is already away from home in the evenings, Tina has to fix her working hours within the timetable imposed by school hours and private after-school child care. Mark does not view his job as particularly adaptable to the children's needs. Though both work for local authorities with quite enlightened 'family friendly policies' (allowing leave for sick dependants, for instance), Mark is reluctant to take advantage of such schemes because it is 'frowned on' for men in management.

Here a spatial arrangement operates, which distances Mark from significant care-giving responsibility. As a consequence, the coordination of this two-job family largely reinforces traditional gender divisions of labour.

Negotiating flexibility: flexibility for whom?

Unravelling the apparent contradictions of global markets and local practices requires that we pay greater attention to the nature and extent of all work undertaken by household members. The relative rigidity, flexibility or uncertainty of spouse working hours and conditions are negotiated over time and space and in the context of a family's life-course stage. This process was made evident in the preceding case by the 'sacrifice' of Tina's career promotion prospects to the 'demand-led' nature of Mark's employment. The biographic narratives suggest that women experience a fluctuating degree of career salience or career ambivalence through the life-course. Female spouse employment strategies aim towards working conditions providing genuine flexibility to the needs of dependants. Louise Livingstone demonstrated this. Career ambivalence, which is generally absent from (or silent in) male employment stories, is evident in several guises: the selection of a short-lived occupation; the selection of an occupation deemed to be flexible to spouse or dependant needs; or the self-imposition of a static occupational position. This latter strategy is described by Mrs Mistry ('dual') as the decision to 'tread water' in her career during a life-stage when her family is her first priority:

> I think one of the reasons I've stayed where I have as long as I have, sort of employment-wise, is because you do have that flexibility. If I moved to another company, I'd have to prove myself over a number of years, so if I needed to go sort of at the drop of a hat to pick my child up, I couldn't.

In contrast, male spouse employment strategies typically serve to militate against increasing job insecurity. In Barking, Mr Lister ('dual') and Mr Livingstone ('flexible') are both training for 'The Knowledge' to become licensed taxicab drivers. In doing so they hope to increase their earning potential by pursuing longer working hours and overtime. Male employment strategies are geared towards occupational mobility (improved earnings), greater security (securing a permanent contract) and better hours (regular hours, shorter hours, preferred shifts). Significantly, geographic mobility is rarely cited as an employment strategy for households with more than one earner.

Any discussion drawing on the popular notion of labour market 'flexibility' immediately begs the question, flexibility for whom? (McRae 1989; Meadows 1996). In a climate of global markets, cultural diversity and shifting gender relations, 'flexibility' assumes a positive connotation – eschewing rigidity and the straitjacket of traditional regimes. But this positive veneer masks ambiguity, ideological bias and inherent contradiction. For employers, flexibility assumes mobility and rapid micro-adjustment to a changing global market. For employees (individual household members), flexibility requires the ability to coordinate household gender divisions of labour within limited time and space parameters (as we saw in the previous chapter). In many ways flexible labour market practices add to the complexity of male and female spouse coordination. Rising numbers of dual-earning households do not correspond with gender role transformation (or wage-earning parity). Instead, female spouse income and career prospects continue to adapt to male spouse employment demands. Households increasingly 'need' two incomes and in cultivating this employment structure they sacrifice individual mobility and household adaptability. What really 'gives' in this exercise of coordination, however, are opportunities for women to combine home and work on an equal footing with men.

The 'flexibility' underpinning contemporary labour market practice takes many different forms. These largely fall under one of two headings. 'Numerical' forms of flexibility make up the 'hire and fire' of part-time and shift work, with employment levels fluctuating according to business demand. State and firm expectations of geographic mobility hinge on the neo-classic economic assumption that labour will 'get on its respective bike' (to paraphrase Norman Tebbit's infamous 1981 speech to the Conservative Party Conference) and move to where new jobs emerge. 'Adaptive' forms of flexibility focus on the expectation that 'core' employees take on new skills and perform multiple tasks according to niche market demands. Here the notion is that 'company men' (supported by versatile wives) as 'portfolio' workers travel freely to deploy skills and services nationally or internationally at short notice. Each form of flexibility presents particular scheduling implications for spouse employment and quite different opportunities and constraints for coordinating social reproduction. Indeed, the diverse nature and extent of employment gathered under the catch-all heading labour market 'flexibility' suggests the term is a misnomer. Take, for example, the impact local labour market restructuring has had on the way Mr and Mrs Morris balance one and a half jobs and care of two young children.

Mr and Mrs Morris: that's all we talk about at work, not seeing enough of our families

Ray and Pam moved to Prestwich from Salford shortly after they were married. Ray has a brother in Prestwich and they saw, through him, that it was a 'nice quiet area' to live – an obvious place to start a family. The couple worked full-time in the first years of their marriage, contributing jointly to the cost of their modest home. Pam took maternity leave with her first child but was made redundant shortly after. She had wanted part-time work but her employer could not offer this, so she accepted redundancy. After a nine-month break, finances got pretty tight. They were struggling to manage on Ray's earnings alone. Pam took a part-time job while her mother looked after the baby, back in her Salford home. This arrangement continued after the birth of a second child. Pam did not think it possible to work full-time because child-care arrangements were complicated. Pam had to drive over to

Salford to drop the children off before going to work and then pick them up again after. This arrangement was now further complicated by the eldest child's attendance at pre-school in Prestwich. Pam has to take full responsibility for child care and the 'school run' because Ray works shifts. Ray continues to work for the same national haulage firm he joined from school. His firm has undergone significant restructuring over this period and four years ago it moved operations from Manchester to Liverpool. This move means that Ray has to commute to Liverpool – an hour each way. Ray resents the fact that his lengthy commute and late-shift pattern reduces the time he can spend with his children:

> At work there's, like, there must be about twenty of us travel up from Manchester and [. . .] that's the main topic of conversation, like, that there's too many late [shifts] and you don't see your family like you should. It does disrupt your family life. I mean, they talk about flexible working all the time, you know, these flexible working practices; it's like their new aim for [haulage firm]. It's how the customers want. The customers want you there within those hours, from the afternoon till evening because that's how they work; it all goes out at tea time. You have to be there at tea time to collect 'em and once you get back to your depot it's like seven, eight o'clock and, you know, that's your finishing time [. . .] We're on rotations, there's one early and they call them middle duties when you finish at like six o'clock, two middles and like seven lates.

Since Ray has been commuting to Liverpool he has been less willing to take on overtime and this puts further pressure on Pam to contribute financially. She complains in turn that her earning potential is greatly limited by the need to be available to pick up the eldest child from school. Ray is unavailable to help because he works late shifts. At no point has the couple contemplated relocation to Liverpool because Prestwich is perceived as a permanent home. Ray describes a move to Liverpool as if it were a foreign place:

> You know, it's a different area, a different like different area aren't they, you know, you've got like on Merseyside and Manchester they're like two separate areas even though they're both in the northwest there's a bit of a divide between the two, if you know what I mean.

Ray is considering switching to a night shift as one way of playing a bigger role in his children's life:

> It would be better paid and the thing is, the reason we was talking about it was I might be able to, say if Pam was working days, it'd sort of work out so I'd be here in the day to, once I'd had me sleep and that, once I'd come home I'd be able to help with the kids more, you know, so I'd be here say in the afternoon when they finish school and at tea time and that.

Pam is not keen on Ray working nights because she believes that they need to spend time together as a couple. There remain only a limited number of ways of balancing residential location, journey to work, employment schedules and child-care arrangements.

Gendered cultures of production, consumption and reproduction

Underpinning this research is the expectation that stories from London and Manchester will each reflect gendered cultures of production, consumption and reproduction. Indeed, evidence of this does permeate the narratives. Yet, surprisingly, common practices and strategies of behaviour frequently appear to circumscribe the 'uniqueness' of place. Superficially at least, households resolve the tensions of home, work and family life in similar ways according to household structure and the nature and extent of spouse employment. Nevertheless, the findings of this project also indicate that households operate from particular positions (of place and cultural identity as well as socio-economic status), which are differentiated and stratified rather than ranged along a single, relative continuum. The biographies suggest that gender divisions of labour (and household employment structures) are contingently articulated through local labour markets. This reinforces our understanding of geographically differentiated gender relations based on secondary data analysis for England, that dual-earning households predominate in the north and north-west whereas male breadwinner households predominate in the south and north-east (Duncan 1991, 1991a; Jarvis 1997).

When we look more closely at 'flexible' households, for instance, it appears that part-time female employment performs a discretely different role in the two urban neighbourhoods. In Prestwich, 'flexible' households consciously articulate a desire to balance the demands of home and work life for both partners. Mr Miliken ('flexible') pronounces his belief that 'time at home with my family is more important than the money in the bank, as long as the bills are paid'. Mrs Mellor similarly evokes a sense of living life to savour the present when she observes that 'we work to live, not live to work'. An apposite vignette of this culture of 'contentment' is suggested by Mrs Miliken when she describes herself and her husband as being 'very slippers and cocoa'. This emphasis on 'balance' contrasts with Barking, where little attempt is made to militate against enduring patriarchal practices. Life here is described as 'fast paced'. Mr Lemon ('traditional') describes the 'industriousness' of his friends and neighbours:

> Everyone here is busy, they are doing something, they are not just out for a jolly, they are out going somewhere, to earn something, to have a meeting, to arrange something. There's no, well I can't see it, there's not a great deal of enjoyment here; there's a great deal of effort expended on making ends meet.

Similarly structured households also pursue different strategies of behaviour in contrasting cities and regions because of the unevenness of housing markets and the impact of this on a 'treadmill' of consumer expectations: the struggle to cope with high housing costs and fulfil dreams of the 'Ideal Homes' interior. Certainly, the role of housing varies by region, location and market position and it was for this reason that neighbourhood selection controlled for income and house price variation.[7] Yet evidence of the diverse role of housing comes across quite clearly. The 'industrious leisure', gender differentiated, home-improvement activities typical of Barking households can be contrasted with the shared activities, home-centred emphasis on 'living for now', enjoying home as a place of retreat from work, in Prestwich households.

Conclusions

An attempt to make the connection between home, work and family life defined the starting point of this research. The key argument is that, if we are to understand the interdependent function of these spheres in contemporary urban contexts, we need to focus attention on the practical concerns of everyday life. We need to explore the inner workings of households, as well as the place-specific context of their reproduction. Moreover, serious attention must be paid to negotiations within households. This serves also as a reminder that processes of cooperation and conflict not only shape individual households but also reinforce wider inequalities between otherwise similar households.

Though to an extent the biographies stress great diversity in the coping strategies of working family households, it is possible to make thematic observations across household employment structures and urban contexts. For instance, patterns of association exist between perceptions of personal security and the ability to sustain current living standards, the proximity and reciprocity of local family networks and the nature and extent of male and female spouse employment. Overall, households with more than one earner are less geographically mobile than single-earner households. This suggests that the negotiation of goals and aspirations for more than one career can result in strategies of inertia or the long-run consolidation of a permanent place of residence. Despite broad patterns of association, stories of the way housing, employment and household structure shape household lived experience are convoluted and fragmented, suggesting that behaviour in these spheres is embedded in a constellation of overlapping and contested interests. Multiple identities are associated with gender, occupational status, housing position, attachment to locale and the interdependent effects of these on residential location and mobility.

A common response to perceived insecurity and increasingly precarious employment (particularly male spouse employment) is for households to maintain a second income, even if low paid, and cultivate a stronger network of social contacts and practical resources. This solution effectively ties them more permanently to a particular local market and feeds into the relative 'rootedness' of low-paid two-earner households. Here, it is not necessarily the presence of a second earner that impinges on mobility, *per se*, rather the presence of a second earner forms part of a strategy of non-mobility. This strategy reflects the consolidation of a permanent place of residence (strong rootedness), in close proximity to local social and kin networks, as a means of combating uncertainty. Great store is set by having multiple sources of income, of keeping overheads low (substituting self-provisioning for paid services), of maintaining reciprocal links and avoiding innovation and change. The price of this strategy appears to be loss of 'family time' and limited opportunity to expand sources of knowledge and information. At a macro level, the implication is that international competition and labour market deregulation contribute to household 'retrenchment'. Global markets, by demanding greater numerical and adaptive flexibility – pushing the proliferation of new part-time, casual and insecure employment – both nourish and incapacitate households with more than one earner. The paradox is that, when we start to understand the local solutions to global processes, it becomes clear that current trends are not sustainable.

While households cite housing costs (the need to accommodate fluctuations in interest rates and maintain owner-occupied housing in good order) as a reason why they 'need' a second income, this 'need' is not determined simply by differential regional house prices or costs of living. It is not the case that higher house prices in Barking, relative to Prestwich, push more 'mothers' into paid employment. Moreover, the availability or shortage of female employment (labour market demand) is rarely cited as a motivating factor in households supporting two earners. What is certainly a factor, however, is the perception of the security and adequacy of male employment as a source of household survival. These perceptions vary socially and culturally, whereby households sharing equivalent material resources (capital and income) differently conclude the 'need' to have a second income.

Housing typically performs the role of a cultural good, a source of self-expression and conspicuous consumption. It is inherently positional, a good that relates directly to socio-economic resources. In the language of welfare and citizenship, housing is a means of determining the quality of children's education (though we saw in Chapters 1 and 4 that the long-distance 'school run' can transcend residential location) and access to wider social, cultural and environmental services. Housing reinforces the relative life-chance of households according to residential location. While it was stressed earlier that the households and neighbourhoods introduced here are not subjects for gentrification research, it is recognised that the residential location strategies of 'ordinary' dual-earning households suggests a certain resonance with a 'break with the suburbs' (Zukin 1987). It is frequently noted that gentrification is a consumption preference specifically associated with high-earning single professional households or couples without children (Munt 1987; Bondi 1991). The findings of this research suggest that a variety of parallel strategies of time–space coordination, such as the strategy of consolidating a place of residence within a hub metropolitan labour market, operate alongside gentrification for a variety of household structures. These strategies of coordination equally demonstrate the interdependence of housing and employment, albeit far less visibly, without capturing the popular imagination in the manner of the 'yuppification' of Georgian terraces in North London.

For the majority of 'flexible' households introduced here, it is not considered possible to raise a family or maintain a home on a single income. Similarly, Pinch and Storey (1991) observed in their study of Southampton households that 'whatever the complex arguments for and against part-time employment, it is playing an important part in maintaining the living standards of families with children' (p. 459). The role of part-time work has changed over the last decade and looks set to continue to change into the future (Hewitt 1996). Clearly, future research needs to expose rather than obscure the role of part-time work and other forms of casual and deregulated employment. 'Flexible' households appear to proliferate in situations where a second income is 'needed' but where full-time female employment is precluded either by the absence of unpaid child care (largely fathers and grandmothers), the prohibitive costs of paid child care relative to earnings or an unwillingness to allow children to be cared for 'outside the home'. Limited child-care options clearly preclude the uniform extension of household employment. It is also the case that paid child care is little help to working parents when their child is ill or if changes in employment schedules clash with day-care opening hours. Greater consideration and imagination needs to be afforded to the coordination of different ways of working and different forms of child care in urban social policy initiatives. New initiatives should be sensitive to the needs of working parents; to facilitate occupational opportunities for mothers in paid employment and an involved role for fathers in the care of their children. 'Family friendly' employment policies, which are all too often interpreted in terms of 'mother friendly' or 'mummy track' policies, need to be reconceptualised as the coordination by men and women of home and work. At present the uncertainty (and fragility) surrounding everyday coordination demands a level of adaptability which frequently exceeds that available from conventional 'breadwinner' employment. For this reason it continues to be women's employment which must be squeezed into day-care opening hours or sacrificed to the needs of a sick child. Everyday social reproduction effectively reinforces persistent gender inequalities in pay and conditions such that we remain a long way from living the 'symmetrical family' forecast by Willmott and Young back in 1973.

Rising numbers of dual-earner households are set to continue. This has implications both for the growing polarisation between these 'work-rich' households and those without any income from paid employment. The further concentration of employment within particular households also has significant implications for preferences articulated through housing choice and location and transport mode and range. More fundamentally perhaps, there are social costs to be addressed. Dual-earner households typically combine long working hours, extend their working hours to compensate for low pay and 'dovetail' spouse employment to accommodate the care of dependants. This effectively limits their access to everyday family life (as poignantly expressed by Mr Morris above) (see also Harrop and Moss 1995; a similar case presented in Hood 1993). Furthermore, it is argued that 'work-rich' dual-earning households are not necessarily increasing their material advantage (though there's no denying that multiple employment is preferable to no employment). To preserve current housing positions, it is frequently perceived as necessary for households to extend the hours worked or the number of earners or both. Peering inside these households also provides a new perspective for the 'social exclusion' debate (Green and Owen 1996 provide a useful overview of 'exclusion', 'disadvantage' and 'deprivation' as popularly applied in current policy debates; Jarvis 1999). The concept of 'exclusion' is generally determined along material lines; unemployment versus employment. What is apparent from this research is that the precarious nature of much employment undertaken by households with more than one earner contributes to unstable and vulnerable household economies.

The 'secrets' behind different and diverse housing and labour market behaviour rest with the very real tensions underpinning everyday household coordination. In practice, flexible labour markets are reproducing remarkably rigid behaviour in these spheres. As a result of the compartmentalisation of policy initiatives concerning housing and employment, attempts made to redress social or economic 'imbalances' in these areas frequently have a countervailing impact (Pratt 1996a). At the same time that further deregulation of the labour market anticipates enhanced labour mobility, the operation of a volatile housing market for owner-occupation, extreme regional house price differential and marginalised public sector provision serve to inhibit the mobility of the households that reproduce labour (Jarvis 1999a). This finding highlights most clearly the inherent contradiction of policy and rhetoric that disregards the reaction and action of local household practices. In effect, time and financial resources required to piece together an increasingly fragmented family life reinforce the imperative that households require two good incomes. At the same time, individual strategies aimed at supporting two competing careers typically reduce household capacity to bring about more integrated lives. The ability to 'go on' with complex employment structures also requires greater recourse to marketised welfare solutions – thus speeding up the treadmill. A growing number of better-paid 'dual-career' households effectively compensate for time spent commuting and long hours away from home by outsourcing domestic reproduction – buying in cleaners, nannies, decorators and ready-to-eat meals. At the other end of this speeding up process, the cleaners, nannies, decorators and caterers have to juggle low-paid servicing jobs in combination with rising housing costs and urban congestion, greatly expanding the role of the hidden gift economy of family and friends. This suggests that one unintended outcome of globalised market restructuring is the marketisation, localisation and casualisation of family welfare.

Notes

1. The concept of the individual or household 'strategy' adopted here is not that which assumes 'rational calculation', 'strategic thinking' or 'information processing', but rather that of the coordination of goals, preferences and uncertainties within the ongoing reproduction of daily practices. The way the concept of household strategies is applied here follows that made popular in recent sociological case-study research (for instance, see Yeandle 1984; Crow 1989; Anderson et al. 1994). The concept of household strategies is in many ways a shorthand for interdependent systems of practices, preferences, goals and aspirations. Strategies are at the same time enabling and constraining, voluntary and structural. Strategies

are typically formed and reproduced from norms and values that are taken for granted. Evidence of the operation of 'strategies' in the biographies suggests support for Hodgson's (1993) understanding that human behaviour can be both rational and subrational at the same time.

2. Historically, Prestwich has been home to Manchester's small but concentrated Jewish community (see Taylor *et al.* 1996 for futher details). None of the households interviewed expressed any particular religious affiliation.

3. Existing literature on east London generally, and Barking in particular, emphasises the key historic function of traditional gender roles and close geographical ties between extended family members (Willmott and Young 1957, 1973; O'Brien and Jones 1996).

4. Average house prices for the two markets point to quite modest differences. In 1997, an 'old' semi-detached house averaged £52,899 in Bury and £67,393 in Barking and Dagenham. This is despite much celebrated divergence in regional house prices (whereby average house prices for Greater London were more than twice those for the north-west region at the last quarter of 1997) (HM Land Registry, Residential Property Price Report, Annual Report, 1998). While housing in Barking is more expensive than the same in Prestwich, the relative cost of living in these areas compensates for this difference. In 1995, the price of the average three-bedroom inter-war (1918–39) semi-detached house was 29 per cent higher in Barking than in Prestwich. At the same time, the cost of living in Barking (allowing for housing costs relating to a mortgage on the above property) was 21 per cent higher than for Prestwich.

5. This data, drawn from the larger research project, is derived from the Sample of Anonymised Records (SARs) of the ONS 1991 Census of Population.

6. This trend corresponds with evidence in existing research on household financial management (Pahl 1984, 1988).

7. A different research project might focus on the regional distribution of household divisions of labour by their association with relative individual income and household earnings levels. Indeed, interest has been expressed in this field (Hills 1995; Hamnett and Cross 1997; Hamnett 1998).

Chapter 6

Towards urban social sustainability

Introduction

This chapter makes the critical connection between the social reproduction of everyday urban life and broader environmental concerns, notably sustainable development. Conventional wisdom views sustainable development as a reconciliation between the goals of environmental protection and economic development. While this is certainly a central relationship, given that sustainable development is a joint challenge across environmental and economic boundaries, we would argue that neither environmental protection nor economic development is an end in itself. On the contrary, each are the means to a larger goal – that associated with the pursuit of a better quality of life for both current and future generations. Mainstream conceptions of sustainable development have a blind spot, seeing environment and economy as a dualism instead of a duality. Accordingly, we need to recover the human/social dimension of sustainability.

In this chapter we argue for the establishment of a new repertoire of tools with which to dig into the generative depth of current environmental problems. We suggest that the dominant capitalist modes of production and consumption need to be re-examined to recover the human scale. Recognition of the duality of socio-economic–environmental relations suggests that this scale is best captured at the interface between individual reproduction and social reproduction. It is urban quality of life which should be placed at the centre of the sustainability debate. Furthermore, given that cities represent both the antithesis of sustainable development and the very opportunities to reverse unsustainable trends, a practical understanding of the meanings of sustainable development requires close observation of the interrelationships between everyday household life and the built environment within cities.

Sustainable development is essentially concerned with key relationships between people and their environments. A full understanding of this interdependence requires a distinction between physical sustainability and social sustainability. They represent the external, physical, aspect and the internal, social, aspect of sustainable development respectively. Physical sustainability is concerned with the relationship between human society and the natural environment as a whole, identifying connections between the capitalist production system and its ultimate material basis. This is what has been widely discussed in mainstream sustainability debates to date. Social sustainability, by contrast, is mainly concerned with the relationships between individual actions and the created environment, or the interconnections between individual life-chances and institutional structures, as outlined in Chapter 3. This is an issue which has been largely neglected in mainstream sustainability debates.

The purpose of this chapter is not to chart all the linked issues of sustainability in the broadest sense, but rather to focus exclusively on the concept of urban social sustainability. Arguably, this context is important on the grounds that it represents the very manifestation of the created environment. Put simply, household decisions concerning where and how to live profoundly shape the profile and quality of the built environment and its associated ecological footprint. Only by reuniting society and nature, bridging the gap between abstract theories of social–environmental sustainability and the concrete dimensions that shape this relationship in practice, is it possible to achieve improved quality of life for current and future generations. Escalating private car dependence, for instance, and associated problems of resource depletion, pollution, congestion and social exclusion have become central concerns at both city and global scales of governance (Girardet 1992; Owens 1992). However, as we have seen in earlier chapters, these issues have much to do with the fragmented location and treatment of employment, housing, education, shopping and other facilities and services. These institutional conflicts, in effect, have a deeper explanation: the expanding logic and utilitarian tendency of industrial capitalism. Using industrialism as the machine and capitalism as the power, capital has changed the world in ways that previous civilisations and natural processes would have taken millennia to achieve. This 'speeding up' is reinforced, in turn, by increased time–space disparity between productive and reproductive activities as illustrated in the narratives introduced in Chapters 4 and 5.

Modern capitalist development is sustained mainly through overcoming the time–space constraints in production and circulation. Nevertheless, the reproduction of individual labour power is by and large a local affair, constrained by distinctly human-scale limitations. Because the reproduction of labour power is a necessary input for the reproduction of the capitalist production system as a whole, the time–space disparities between productive and reproductive activities become a major obstacle to sustainable development in capitalist society. Accordingly, it is essential to explore the time–space relations between productive and reproductive activities in order to gain a practical understanding of the meanings of sustainable development. This is the key to a holistic approach to planning for sustainable development. At the centre of such a holistic approach is the totality of everyday urban life. In the light of its repetitive nature and its collective effect, this micro aspect of time–space coordination represents the crystallisation of different institutional connections at higher levels.

The remainder of this chapter develops and connects each of the ideas introduced above in turn. First, the current arguments in mainstream sustainability debates are mapped out in summary by way of background. This brief review serves to highlight a critical gap in understanding that exists between theory and practice. Discussion moves on to the *meaning* of sustainable development and the need to make a clear distinction between physical sustainability and social sustainability. Attention then turns to the generative depth of 'unsustainability', tracing the social origins of environmental problems to the underlying logic of industrial capitalism. This discussion is fleshed out with concrete reference to the practical implications of the British government's planning policy for sustainable development. Finally, we conclude this chapter by arguing for a need to establish an understanding of 'livelihood science' to provide an integrated approach to planning for sustainable development.

The sustainability debates

The term 'sustainable development', or simply 'sustainability', has become popular since publication of the Brundtland Report (see WCED 1987). Conceptually, it stresses the importance of intergenerational equity; empirically, it stresses the need for a simultaneous achievement of both developmental and environmental goals (Cleveland 1987; Redclift 1987). Since Brundtland, this concept has been applied to an increasingly broad range of issues and spawned a number of competing interpretations (Pearce *et al.* 1989; Holding and Tate

1996). Arguably, the precise meaning of sustainability is largely concealed by ambiguous and abstract terminology.

Sustainable development: a complex issue

Most sustainability definitions follow on from the Brundtland Report. For instance, sustainable development builds on the notion of 'development which meets the needs of the present without compromising the ability of future generations to meet their own needs' (WCED 1987: 43). This definition has been criticised, however, for being 'too vague' (Bartelmus 1994: 69), more like 'a device for mobilising opinion rather than as an analytical concept for developing specific policies' (Blowers 1993b: 5). Indeed, the Brundtland Report itself applies the concept inconsistently (Pearce *et al.* 1989: xiv). On the one hand, a great majority of the sustainability discussions focus on the biophysical environment that comprises the Earth's life-support system. On the other hand, however, an increasing number of sustainability discussions now stress the importance of the socio-economic environment, encompassing people and their activities, which both shape and transform the biophysical environment. Concepts such as social sustainability, economic sustainability, community sustainability and cultural sustainability are now considered to be important elements of sustainable development (Mitlin and Satterthwaite 1996: 25). This suggests that the nature of sustainable development is both complex and dynamic. One of the major contributions of the Brundtland Report is that it rightly addresses the intrinsic links between economic and environmental issues and forcefully criticises current policy responses which compartmentalise institutional decisions (WCED 1987: 310–12). Still unclear, however, is the true meaning of sustainable development and its practical implications.

In terms of theoretical foundation, one of the most popular approaches to sustainability issues remains that of neo-classical economics. This emphasises the allocative role of the price mechanism and how this can contribute to 'efficient' resource use, a view represented by the *Blueprint* series[1] (see Pearce *et al.* 1989; Pearce 1991, 1993, 1995). The main neo-classical economic argument is that unsustainability problems arise because environmental services are not sufficiently or completely valued in the processes of current economic decision-making. The price mechanism has wrongly recorded environmental goods and services as having zero or very low costs, so the economic system tends to overuse these undervalued environmental resources.

Although some environmental costs can be assigned to the use of scarce or vulnerable resources, the major weakness of the neo-classical economic approach is that of economic reductionism. While neo-classical environmental economists rightly note that 'development' should not be conflated with 'growth', they fail to recognise that economy is only a part, though a very important part, of society. Actions based on this conception effectively widen, rather than reduce, the inequalities between regions and between social groups (Mehmet 1995: 125). Some authors go further, to argue that environmental crises cannot be resolved by economic measures alone because the crises are effectively intrinsic to the subject of economics itself (Jacobs 1991). It is the overly emphasised economic logic that leads to 'development' being unsustainable. Accordingly, what is urgently needed is, first, closer examination of the essential links between environmental and economic crises and, second, development of an integrated approach that can establish a common ground between environmental and economic goals.

Sustainable development: a reconceptualisation

Mainstream conceptions of sustainable development tend to stress either environmental sustainability (strong sustainability) or economic sustainability (weak sustainability) (Turner *et al.* 1994: 31). At the same time, they share a common blind spot: they pay remarkably little attention to people themselves, their thought and action. This is surprising given the way both environment and development are effectively colonised by human endeavour. As the Brundtland Report notes, 'the "environment" is where we all live; and "development" is what we all do in an attempt to improve our lot within that abode' (WCED 1987: xi). In other words, sustainable development is not only concerned with environmental protection and economic development *per se*, but also concerned with the very interdependence of people and their environments. The human/social dimension of sustainability needs to be made equally as explicit as socio-environmental and socio-economic relations.

Central to reconceptualising the meaning of sustainable development is an understanding that this interdependence of people and their environments effectively constitutes a duality of structure. According to this view, environment, development and people should not be seen as discrete entities, as a dualism. Rather, they represent an interdependent whole, a duality of people's livelihoods and their environments. To explore this notion further, we need extend our understanding of each of these overlapping spheres.

The concept of environment: natural environment and human-made environment

While 'the environment' is central to sustainable development debates, the orthodox conception is that of a 'natural environment' about which human activity intervenes. This natural environment is viewed as a given, pre-existing to human beings and external to human societies. It is a realm of impersonal object, to be studied, conquered and explored by humans (see Fuller 1988; Dickens 1992, 1996). Such views represent an alienation between humans and nature: society is not only a part of nature, but is also apart from it. Clearly, there are pre-existing conditions of the natural environment, such as the atmosphere, oceans, lands, and the ecosystems as a whole, of which human society is an integral part. However, environment has another dimension that cannot be separated from where we live and what we make of nature – that which denotes human surroundings and their milieux. It is this situatedness that makes the environment relevant to our survival and development (Harvey 1996: 116). This means we need to closely consider the modified conditions of the built environment. Here, people do not merely adapt to an external material world. Rather, they seek to master their environments: the very idea of the created environment.

After several million years of hunting and gathering and several thousand years of tilling the soil, human society has entered a new era in which the created environment plays a dominant role. In the last two centuries, fast and large-scale urbanisation and dramatic socio-economic changes suggest that few areas of the globe can escape intended or unintended human intervention. In the UK, for example, very little true wilderness remains. Only the vestiges of once great natural forests remain in places like Sherwood, the New Forest and the Forest of Dean (McCormick 1995: 161). The 'wilderness' in national parks and conservation areas, the 'greens' in farmlands and the forests, both are the product of purposeful human intervention, not to mention the 'grey jungles' constituted by skyscrapers and dense networks of roads and rails in the cities. To quote Beck (1992: 80), this implies that

> At the end of the twentieth century, it means the end of the antithesis between nature and society. That means that nature can no longer be understood outside of society, or society outside of nature. The social theories, which understood nature as something given, ascribed, to be subdued, and therefore always as something opposing us, alien to us, as non-society, have been nullified by the industrialisation process itself . . . At the end of the twentieth century, nature is neither given nor ascribed, but has instead become a historical product, the interior furnishings of the civilisational world, destroyed or endangered in the natural conditions of its repro-duction. But that means that the destruction of nature, integrated into the univer-sal circulation of industrial production, ceases to be 'mere' destruction of nature and becomes an integral component of the social, political and economic dynamic.

In short, we no longer stand in direct contact with nature, but live in a mediated and manufactured space (Bertilsson 1984: 48). The food we eat and the tools we use are increasingly produced in places further afield and at points in time long before actual consumption. At the extreme, in a highly artificial world, the 'natural' environment would become an integral part of the created environment, such as in a spaceship. In other words, while people seek to modify their environments to accommodate their needs, they also change their behaviour to adapt to the modified environments. Accordingly, to understand the meaning of sustainable development we must address the importance of the created environment, as well as its relationship with the natural environment.

The concept of development: economic development and socio-economic development

The relationship between man-made and 'natural' environments can be ex-plored with reference to the role of environmental services in the fulfilment of developmental needs. A distinction is often made in this regard between the quantitative and the qualitative dimensions of development, or between growth and development (see Redclift 1987; Pearce *et al.* 1989; Turner *et al.* 1994). Economic growth is at best a means to an end, but not an end in itself (UNDP 1992: 2; World Bank 1992: 34). Development in the fuller sense, by contrast, is not only judged by the production of goods and services (that is, the accumulation of wealth) but also by reproduction considerations, such as improvement in the quality of life. The latter involve questions of distribution and positioning. Some human assets (labour power) can be owned, traded and thus valued in the market as produced assets, but human well-being cannot be 'summed up' or 'averaged out' as the monetary evaluations of commodities and productions. Following this principle, sustainable development needs, in practice, to transcend the narrowly defined notion of economic growth and reach a wider socio-economic dynamic of a comprehensive form of development.

Greater understanding of the relationship between production and reproduction and their roles in the aspiration for development helps clarify the socio-economic dynamics of development. In short, production is concerned with the transformation of materials, an input–output relation (Smith and O'Keefe 1996: 288). By contrast, reproduction is concerned with the transformation of being. It has a twofold meaning. On the one hand, reproduction represents a momentary happening, a single event such as consumption. In Marxist terms, it is the 'reproduction of labour power', a realm of individual reproduction. On the other hand, reproduction also represents a cyclical or repetitive process relating to the continuity of the production system, that is, re-production. In this way it belongs to the realm of social reproduction.

As might be expected, consumption must be supported by the production of goods and services, and the reproduction of the production system relies on continuing consumption. The relationship between production and reproduction, accordingly, can be understood in two ways. One is a compositional dualism: as production and consumption. Another is a constitutional duality: as production and re-production. Progress in the transformation of materials, or the reproduction of the production system, is defined as productive development. Improvement in the transformation of being, or the reproduction of labour power, is defined as reproductive development. The former is so-called economic development, it is a means to the end of development. The latter can be termed socio-economic development and constitutes the ultimate goal of development.

Yet, the difference between productive and reproductive development is more than a contrast between quantitative and qualitative development. A sustainable overall development needs an adequate channelling between productive development and reproductive development in the dualities of production, consumption and re-production.

Strictly speaking, at the start of the twenty-first century there is no environment that is completely 'natural' or 'artificial'; it is just a question of degree. Boundaries between natural and built environments are increasingly blurred. On the one hand, the natural environment is increasingly shaped by human intervention; on the other hand, human activities are increasingly conditioned not only by natural processes but also by what people have made of nature. In other words, 'unsustainability' issues are, in essence, social problems, problems created by, and eventually impacting on, people themselves (Commoner 1973: 23; Beck 1992: 81). As Singh (1989: 155) argues, 'when any environmental issue is pursued to its origins, it reveals . . . that the root cause of the crisis is not to be found in how men interact with nature, but in how they interact with each other.' Modern civilisation is established on the appropriation of nature via a relationship of transformation between human and environment in a continual process of production and reproduction. In order to address the interconnections between environment and development, it is necessary to draw out the role of human mediation, in particular the intrinsic links between individual actions and social structures.

The concept of people: individuals and society

The relationship between individuals and society, or between agency and structure, has long been the focus of sociological study, as outlined in Chapter 2. There are traditionally several contrasting views here. First, for instance, the voluntarist camp, represented by Weber, argues that society is constituted by individuals and their intentional behaviour. In contrast, the reificationist camp, represented by Durkheim, argues that society has a life of its own, external to and coercing individual behaviour. The dialectical camp, developed by Berger and his associates, argues that society and individuals have a dialectical relationship: individuals create society and society produces individuals in a continuous dialectic reproduction (for a brief summary, see Gregory 1981; Bhaskar 1989; Walmsley and Lewis 1993).

Though contradictory at a philosophical level, these three theories share a common problem: they all subscribe to the dualism between individual behaviour and social structure. A fourth camp, which aims to resolve this dualism, is the structurationist camp, represented by Giddens and his theory of structuration. Giddens (1984: 25) argues that

> The constitution of agents and structures are not two independently given sets of phenomena, a dualism, but represent a duality . . . the structural properties of social systems are both medium and outcome of the practices they [agents] recursively organise.

This view sees individual actions and social structures as the two sides of the same duality coin. Human actions create the structures of society and these structures provide the contexts for social interactions. In turn, human actions reflect and recreate these social structures. In Bhaskar's terms, 'Society is both the ever-present *condition* and the continually reproduced *outcome* of human agency' (Bhaskar 1989: 34–5, original emphasis). Arguably, the duality relationship between human agency and social structure is the binding element required to link environmental and developmental goals in a sustainable harmony between people and their environments.

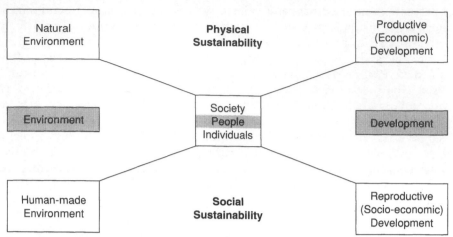

Figure 6.1   The relationship between physical and social sustainability

Physical sustainability and social sustainability: external and internal dimensions of a sustainable overall development

Having argued that the concepts of environment, development and people cannot be reduced to the narrowly defined notions of the natural environment, economic development and human society, it is important to stress that duality between people and their environments can be explored from two perspectives: physical sustainability and social sustainability. Physical sustainability is mainly concerned with the material interconnections between economic development and the natural environment. Social sustainability, by contrast, is mainly concerned with the social connections between socio-economic development and the built environment. They constitute the external and the internal dimensions of a sustainable overall development. The issues of physical sustainability are well rehearsed, though hotly contested, in mainstream sustainability debates. Issues of social sustainability, however, are neglected and largely misunderstood. Figure 6.1 illustrates this relationship between physical sustainability and social sustainability. It also indicates the reconceptualisation of a sustainable overall development.

In a sense, the distinction between physical sustainability and social sustainability is similar to Ward's conception of sustainable development, that is, meeting the 'inner limits' of human needs and rights without exceeding the 'outer limits' of the planet's ability to sustain life now and in the future (Ward 1976). Viewed this way, social sustainability is both the prerequisite and ultimate goal of physical sustainability. Yet conventional sustainability debates tend to focus on the external, physical aspect of sustainability with little regard to the internal, social aspect of sustainability (see van den Bergh and van den Straaten 1994). Attention is limited to problems of sub-sustainability and pseudo-sustainability. The former sees social sustainability as an extension of economic and environmental sustainability. Sustainable development is considered within, rather than between, sectoral and regional boundaries in a competitive and potentially divisive model of social–environmental interaction. The latter is concerned simply with the material conditions of a society's production system. Here the goal of sustainability is likely to be achieved at the expense of highly geographically uneven opportunities for social reproduction.

In short, interpretations of sustainability that focus exclusively on the physical maintenance of the natural environment fail to take account of key underlying welfare considerations. Neither sub-sustainability nor pseudo-sustainability can be viewed as being at all sustainable in this regard. A 'sustainable overall development', accordingly, requires that equal attention be paid to both physical and social sustainability. If we agree that unsustainable development is, in essence, a social problem then we must have a full account of the internal, social origins of both environmental and economic problems.

Industrial capitalism and social sustainability

It is widely accepted that the origins of current environmental problems must be traced back to dominant modes of production and consumption. This topic of discussion provides a popular point of departure for sustainability debates (Johnston 1989; Robertson 1989; Singh 1989). Following this trend, it could be argued that unsustainable development is in effect a 'Western problem', a problem closely associated with the process of Western modernisation. This is not to say that problems of non-sustainability only exist in the Western world. Rather, given that the world system as a whole is moving towards a Western style of development, that is, an industrial market economy, promoting Western democracy and the like, it means that this ideology of development spreads out from its roots according to the historical geography of Western capitalism. This is characterised by the aspiration of 'becoming developed' in many Third World countries. While arguing that social sustainability is both the premise and the ultimate goal of a physically sustainable development, we need to look to the logic of industrial capitalism for explanations why this critical balance has been so neglected.

Industrial capitalism: a growth machine

At the heart of Western industrial capitalism is combined the utilitarian tendency of industrialism and the expanding logic of capitalism in the appropriation of natural resources and the transformation of social systems. Aided by processes of urbanisation, imperialism and colonialism, industrial capitalism has spread across the world on an unprecedented scale over the last two centuries (see Brown 1974; Blaut 1975; Mandel 1975; Etherington 1984). Yet, industrialism and capitalism are two distinct systems that should not be conflated, either conceptually or empirically (Giddens 1985: 123). On the one hand, industrialism essentially refers to the transformation of the means of production. This is characterised by the process of industrialisation that alters the human–nature relations via large-scale mechanisation and high energy consumption (especially non-renewable fossil fuels) in both production and everyday life (Giddens 1990: 60). It denotes a utilitarian philosophy, illustrating how human societies use the results of scientific–technological advances to meet the needs of developmental aspirations, mainly through the appropriation of nature. On the other hand, capitalism is the power preceding industrial development, providing much of the impetus to its emergence (Giddens 1990: 61). In other words, capitalism is the driving force that has extended the arena of Western industrialisation to a truly global scale. It refers to a historically specific form of economic and social organisation in which (a) the direct producer is separated from ownership of the means of production and the product of labour process; and where (b) this separation is effected through the transformation of labour power into a commodity to be bought and sold in the labour market regulated by price signals (Gregory 1994: 40–1). Of course, industrialism could not function on a local scale without capitalism. Nevertheless, the relationship is symbiotic rather than mutually exclusive. Capitalism has facilitated, and has been reinforced by, the separation between production and reproduction. This separation, in turn, has resulted in the alienation of humans from nature.

Traditionally there are two main strands of Marxist-influenced political–economic critiques on industrial capitalism: theories of class conflict and those of capital accumulation (see Gottdiener 1985, ch. 3). The collapse of Marxist-Leninist socialism in Eastern Europe has nullified the thesis of class conflict for some commentators and has given rise to a dangerous self-congratulation in the West. This is too often seen as a victory of capitalism. The collapse of socialism has effectively given industrial capitalism more space to expand, a logic that is principally geared towards the search for surplus value. As Hill (1977) notes:

> Capital accumulation, the production of surplus value, is the driving force of capitalist society. By its very nature, capital accumulation necessitates expansion of the means of production, expansion of the size of wage labour force, expansion of circulation activity as more products become commodities. (Hill 1977, cited in Gottdiener 1985: 87)

To quote Fukuyama (1992), capitalist expansion is 'accumulation without end', or what Saunders (1995) calls 'the growth machine'. Industrial capitalism is like 'a monster created by Frankenstein, something powerful, out of control, destructive and seemingly uncontainable' (Saunders 1995: 54). Accordingly, knowledge of the practical manifestations of this expanding tendency opens up a useful route to understanding the concept of social sustainability.

Uneven development between production and reproduction in space and time

The consequences of capitalist expansion are traditionally conceived as uneven development between cores and peripheries at global, regional and local scales of analysis (see Webber 1982; Harris 1983; Massey 1984; Smith 1990).

Attention typically focuses on the productive side of development, emphasising different degrees of capital accumulation and their role in the process of capitalist production. Because these analyses ignore the reproductive side of development, they only sketch the surface of unsustainable development issues, leaving their underlying causalities almost untouched. In contrast, social sustainability is concerned more specifically with uneven development between productive and reproductive activities, an issue embedded in the relationship between economic and environmental considerations and between human agency and social structure. Unevenness between productive and reproductive development can be illustrated by the existence of time–space tension between capitalist modes of production and consumption.

Industrial capitalism has effectively brought about a radical institutional change in the organisation of everyday life, by separating productive and reproductive activities across both space and time. The implications of such a change can be illustrated by comparing pre-industrial with industrial and post-industrial social systems. In a hunting and gathering society, for example, production and reproduction are typically one and the same thing. Consumption takes place in close proximity to the process of production. In this arrangement there is little sense of uneven development between production and reproduction, except in the general manifestation of 'underdevelopment'. In agrarian society, production and reproduction take place in a relatively limited series of time–space zones, closely tied to the land. Although in this stage of development trade relations increasingly link cities with rural areas, the constraints in production, food conservation and transportation typically limit the possibilities of time–space expansion in both production and reproduction. Even the 'urban' inhabitants both live and work within the city walls. In other words, both production and reproduction are localised activities in agrarian society.

With industrial capitalist society, activities of production (especially mass manufacturing) and reproduction increasingly function at different places and times. This results in a separation between workplace and living place, as well as the associated division between work time and leisure time. Most importantly, the time–space disparities between production and reproduction apply not only to the working class, who sell their time and labour in the market, but also to the capitalists, who own the means of production. This separation in turn serves to reproduce an economic entity geared towards the search for profit: the capitalist production system. Families are no longer the only units of production. On the contrary, firms owned by public stockholders play a dominant role in capital accumulation. Marxists tend to explain the potential conflicts between production and reproduction as a class conflict. Yet a more convincing argument is that this represents an *institutional conflict*. With the development of modern security markets, the means of production is increasingly owned by the public, including the workers, rather than it being in the hands of a small number of capitalists. Thus, uneven development is a conflict shared by workers and capitalists. Arguably, this institutional conflict is closely related to the crisis of unsustainable development.

In *A Contemporary Critique of Historical Materialism*, Giddens uses the concept of time–space 'distanciation' to explain the driving principle of modernity, illustrating how societies are 'stretched' over shorter or longer spans of space and time (Giddens 1981: 90). In this view, productive development is based on its ability to stretch over wider space and longer time. Capitalist expansion, accordingly, can be understood as a manifestation of time–space distanciation.

In the realm of reproduction, however, 'time–space co-presence' is as important as 'time–space distanciation'. Activities as simple as eating, sheltering and wearing clothes are matters of 'here and now' that can hardly be deferred or undertaken from afield. In other words, the reproduction of labour power is the internal constraint for an unlimited expansion of capital while the reproduction of the production system acts as the external constraint of development. Again, this can be demonstrated by a historical review.

In primary and agrarian societies, social systems are organised by activities with a higher degree of presence-availability in both production and reproduction. Therefore, the issue of uneven development is less marked, although there exists the problem of underdevelopment. In capitalist society, by contrast, productive development reaches a new high: more and better products are manufactured and circulated on a truly global scale. With the help of scientific and technological advances, including progresses in transportation and communication, industrial capitalism increases its rates of circulation by stretching over space (creating new markets) and time (reducing or extending product life cycles). Industrial production has increased 50 times in the last one hundred years and increased 7 times since 1950 (WCED 1987: 15; Saunders 1995: 56). Industrial capitalist society, in Harvey's words, is built on the principle of 'accumulation for accumulation's sake, production for production's sake'. This explains why economic performance, economic growth and economic expansion are prone to be the abiding interest, if not the obsession, of all modern societies.

Reproductive activities, on the other hand, by and large remain as local affairs by virtue of the time–space constraint of embodiment. This implies a fundamental gap between globalised production system and localised reproduction practices. The inherent problems in capitalist society are overaccumulation of capital and overproduction of commodities. The way to avoid the predicaments of overaccumulation and overproduction is either through an extension of the loops of capital circulation, by directing capital into tertiary sectors, or through the process of 'creative destruction', by accelerating commodities' turnover time (Harvey 1989b: 27). In other words, what industrial capitalism does best – stimulating expansion – also signals its greatest weakness. It is like a candle burning at both ends. On the one hand, expansion in production and consumption requires more resource inputs. But the replacement rates of environmental services and natural processes are unable to keep up with the rates of capital accumulation. On the other hand, the expansion of the production system as a whole does not necessarily imply that everybody can share this growth in consumption to the same extent. This typically disadvantages those whose needs are not properly reflected in the market as effective demands. In other words, industrial capitalism serves to undermine its own foundation through endless expansion. Accordingly, the ability to bridge the time–space friction between production and reproduction, and between individual reproduction and social reproduction, forms an essential part of a sustainable social development.

Institutional conflict and social sustainability

The time–space disparities between production and reproduction have profound implications for the organisation of both everyday life and the overall social structure. For example, when discussing the issue of urban settlement, Ward (1975: 39) argues that 'the very word "settlement" is in some measure a contradiction. In many ways modern man is living with "unsettlement".' Haughton and Hunter (1994: 9) add: 'our places of work, recreation and residence all differ and change over time, so that . . . we are always on the move within and between our cities.' In other words, such time–space disparities necessarily create a 'mobile society' that demands a large amount of movements in space and time.

By virtue of their routinised and repetitive nature, the growing scale of motorised journeys for work, schooling, shopping and leisure purposes, in particular those trips made by private car, have profound environmental, economic and social implications. A tiny time–space friction between different daily locales, such as between the places of home and work, may create a large volume of travel. The growing scale of transport, in turn, creates a range of problems, such as the waste of non-renewable fossil energy, the generation of different kinds of pollution, road casualties, and the associated problem of social exclusion for those who have poor access to suitable transport. These issues represent the very antithesis of sustainable development (see, for example, Newman and Kenworthy 1989; CEC 1990; Breheny 1992a; DoE 1994a; Anderson *et al.* 1996). In other words, many problems associated with unsustainable development and practice are deeply rooted in the institutional conflict created by the expanding tendency of industrial capitalism. While we are dealing with the external, physical dimension of sustainability, we also have to take a full account of the internal, social aspect of sustainability.

Restoring the human scale of sustainable development

Lack of attention to the social aspect of sustainability is an inherent weakness of mainstream sustainability debates. It is a major obstacle to the pursuit of an 'overall sustainability' in practice. This is why it is so important to dig deep into the underlying logic of industrial capitalism and its time–space implications for individual reproduction and social reproduction as the basis for reconceptualising sustainable development. Indeed, a significant body of literature exists to question the adequacy of capitalist expansion (see, for instance, Goldsmith *et al.* 1972; Meadows *et al.* 1972; Ward and Dubos 1972). Schumacher's (1973) seminal text *Small Is Beautiful*, for instance, forcefully criticises the capitalist mode of production and its emphasis on growth. Schumacher calls for a restoration of the 'human scale' to institutions and processes. These ideas have recently gained ground in popular espousal of the 'compact city' and 'urban village' ethos. The argument is that modern urbanisation should return to the European tradition of mediaeval cities by concentrating people and activities in the confined areas of existing cities so that the need to travel can be minimised. Presumably, this will reduce adverse impacts on the environment (CEC 1990, 1993; Elkin *et al.* 1991; Sherlock 1991). The prospects of this planning strategy succeeding are highly debatable (see, for example, Beckerman 1995; Douthwaite 1986). Restoration of the human scale does not necessarily imply an unthinking U-turn to small-scale, localised types of development, as suggested in the Small-Is-Beautiful argument.

Arguably what is more important for the restoration of the human scale of development is to 're-embed' the 'disembedded' practices into concrete space and time. The crises of unsustainable development and industrial capitalism are not simply a question of scale, but a more fundamental issue of time–space channelling between productive and reproductive activities in the duality between individual reproduction and social reproduction. The missing link constitutes that which connects the routinised practices of everyday life with institutionalised social structures in the world's cities. An urban focus is justified on several grounds. First and foremost, many critics argue that cities represent the very antithesis of sustainable development: cities are net consumers of natural resources and major producers of pollution and waste (Girardet 1992: 86; Owens 1992: 79). They are regionally unsustainable. Moreover, cities are largely associated with crime, vandalism, deprivation, unemployment and many other socio-economic problems, including deteriorating infrastructure, inner-city decay and community disintegration. They are therefore also frequently socially unsustainable.

Second, cities are the dominant forums of modern civilisation. They are the foci of production, distribution, exchange and consumption, where the heart of the capitalist mode of production lies (Johnston 1989). The world as a whole is becoming economically, culturally and environmentally more interconnected. Cities are the nodal locations where unprecedented flows of resources, wastes, information, traded products and services, and financial capital and labour concentrate. Given that the sizes of both cities and urban populations have been growing consistently over the years, the issue of making cities sustainable becomes one of the greatest challenges facing both developed and developing countries (WCED 1987: 235). More than ever before, the world as a whole is moving towards an urban society as a result of the fast growth of Third World cities. The concentration of people and activities in cities provides greater latitude for policy intervention. It is politically difficult to prioritise global sustainability in the face of local unemployment, poverty, poor housing, deteriorating infrastructure and so on. Therefore, to resolve existing urban questions by linking the embedded daily practices with the disembedded institutional structures in cities is to pursue sustainable development from within.

Most importantly, cities are built by and for people. They are more than the static structures of bricks and mortar; they are also the key settings of social transformation. Urban spaces carry the very spirit of the created environment in terms of their massive scale, versatile functions and, most importantly, their role in social transformation. Cities are the major locations of social, economic, political and cultural interactions. Today, where we work, rest or play depends more on the spaces and social relations that we have created or modified than it does the natural or inherent characteristics of different locations. Accordingly, the urban focus of the created environment provides a *meso* context of socio-spatial analysis through which the duality of structure can be understood more easily. To facilitate this practical under-standing, and to establish the connections between everyday life and a broader environment concern, it is useful to reflect on the British government's sustainability strategy as a real-life example. The aim of this section, then, is to demonstrate why this human scale is so important for an effective pursuit of the goal of sustainability.

Spatial integration: planning for sustainability in the UK

As discussed in Chapter 3, London is, in effect, a divided city in terms of the time–space relations between different institutional structures as well as between individual household lives and the institutional structures. Similar problems also exist in other British cities, though the nature and extent of institutional disparities in time–space vary from one city to another. Clearly, efforts have been made to develop an integrated, holistic approach to deal with these issues. Most importantly, this goal is considered in conjunction with that of planning for sustainable development (see DoE 1990, 1992e; Blowers 1993a; Jacobs 1993; DoE/DoT 1994; Healey and Shaw 1994; Myerson and Rydin 1994; Buckingham-Hatfield and Evans 1996). As a general rule, the British government's sustainability policy can be characterised as a strategy of 'spatial integration' that echoes the 'compact city' proposal introduced above. Although there is consensus concerning the need for an integration of planning policies, two problems stand out in the British case: first, promotion of the environment as an overriding goal in planning for sustainability and, second, the ways that planning policies are integrated. Most importantly, an inherent problem of this strategy is the fallacy of a 'nominal approach'. This approach sets out to change the surface structures of urban land use but gives little consideration to the deeper relations between institutional structures. It could be argued that the potential success of the 'environmental turn' in British planning is limited by its failure to address the intrinsic interdepend-ency between individual household lives and the created environment.

Urban reconcentration and mixed-use development: concentration or integration?

The central theme of the British government's sustainability strategy is a policy of 'urban reconcentration', that is, to direct various developments back to existing urban areas and to make the most efficient use of urban space. The underlying assumption of this concentration strategy is that, by virtue of their density, towns and cities provide more efficient ways for organising people's daily activities and consequently reduce the need to travel. In order to achieve this goal, it adopts a policy of mixed-use development around key nodal locations that are accessible by a range of transport, in particular those environment-friendly modes of transport such as pedestrian, bicycle and mass transit. In other words, the British government's sustainability strategy is to use the existing urban areas in the most effective ways by making them more attractive places to live and work.

In order to translate the policies of urban reconcentration and mixed-use development into substantive planning goals, in particular via the preparation of development plans, the British government has, since 1992, revised the Planning Policy Guidance notes (PPGs) to reflect sustainability concerns. The central theme of the latest revised PPGs is, in short, an integrated approach that stresses the coordination between transport and land-use planning. Of course, transportation functions in two competing ways. On the one hand, an effective transport system is vital for both local and national economies. On the other hand, traffic growth represents a major threat to the environment. It is not surprising that the British government's sustainability strategy is focused on reducing the need to travel, including reducing the length and the number of motorised journeys, reducing car dependence and encouraging alternative means of transport with less adverse environmental impacts. But these goals cannot be achieved by transport policies alone; supportive policies in other sectors are essential inputs too.

The transport policies set out in PPG 3 seek to make best use of current transport infrastructure and to encourage the shift of transport modes from the private car to public transport and other environment-friendly modes of transport. Most importantly, these transport policies are seen as an integrated part of the sustainability strategy that aims to direct assorted developments back into existing urban areas, in particular at or near nodal locations that are easily accessible by different modes of transport (DoE/DoT 1994).

To achieve this goal, the revised PPG 4 emphasises that jobs and homes should be accessible to each other over large parts of urban areas (DoE 1992d). It argues that a broader economic base with an adequate balance between manufacturing and service industries within urban boundaries will secure long-term prosperity and thereby maintain a sustainable relationship between homes and jobs. Accordingly, while strengthening Central London's status as a centre of finance, commerce, culture and tourism, a particular objective for industrial and commercial development is to broaden London's economic base by attracting manufacturing industries to move into east and/or Outer London.

However, a balanced economy with less need to travel must be supported by the housing sector. Given that housing demand has always been high in London, even at times of recession, the British government adopts a strategy of maximising housing provision in London. But the land supply for new housing development is very limited here, partly because the government is reluctant to release land from green belt sites. Priority is therefore given to recycling urban land, in particular run-down, vacant or derelict sites. Otherwise, housing development that is not occurring at or near key nodal locations will be required to take into account the provision of necessary infrastructure and to take advantage of the least congested parts of the transport network (DoE 1994b).

As far as retail development is concerned, the British government tries to encourage in-town provision. It argues that retail development should 'sustain or enhance the vitality and viability of town centres . . . and ensure the availability of a wide range of shopping opportunities to which people have easy access' (DoE 1996b, para. 1.1). Accordingly, the British government adopts a sequential approach to the selection of suitable sites for new retail development: the first preference is for town centre sites, followed by edge-of-centre sites, district and local centres and only then out-of-town sites in locations that are accessible by a choice of means of transport. In view of the fact that London has a dense network of town centres with good public transport links, it is believed that the existing town centres should continue to be the major locations of shopping provision and other facilities and services so that one trip can serve several purposes, in particular for those who do not have a car (DoE 1994b; GOL 1995).

Diversity, flexibility and accessibility: a socially sustainable city

There is a general consensus that cutting the numbers and distances of motorised journeys is important for the goals of environmental protection, economic development and social justice. To do this requires an integrated, holistic approach. However, the British government's sustainability strategy of urban reconcentration, mixed-use development and land use/transport coordination is problematic on several grounds.

First, the decision to direct activities and developments back into towns and cities may reduce the number of car trips, especially those trips made between and within suburbs, but efforts to make city centres strong in every aspect, as centres for offices, shopping, entertainment and housing, can be as problematic as unchecked decentralisation. At its best, the rejuvenation of the central areas may attract more people travelling from areas further afield to work, shop, entertain, educate and the like, although many such trips can be served by public transport. As a consequence, it will create extra pressure for urban sprawl. At the worst, the movement of jobs to the urban fringes (particularly those with fewer skill requirements) – policies aimed at concentrating dwellings at high densities – may result in reduced living conditions, longer and more difficult journeys to work and higher rates of unemployment in the inner cities.

It has been argued that the dispersal of employment, residence, retailing and other facilities and services actually reduces, rather than increases, the numbers and distances of journeys. People stop making longer suburb-to-city trips, but instead make shorter suburb-to-suburb trips (Gordon and Richardson 1989a, 1989b; Gordon et al. 1991; Brotchie et al. 1995). However, as the London case illustrates, the problem is that considerable proportions of such trips are made by car. It could be argued that what is at issue is not the spatial configurations of development *per se*, that is, a choice between concentration and decentralisation. Rather, what is more important in the British government's sustainability strategy of spatial integration, or the idea of the 'compact city' in general, is its practical implications for the coordination of everyday life.

It is naive to assume that the need to travel can be largely reduced when various kinds of developments and activities are concentrated in the cities. It may be the case for smaller towns and cities because smaller scales of development and simpler relations between institutions and people make it easier to maintain a self-contained condition. But it is very difficult for large cities to maintain such a self-sufficient pattern of development in confined urban boundaries. London is a good example in this regard. In the processes of modern urbanisation, what has changed is not only the size of population and the scale of activity, but also the very nature of social relations: a more complex web of social relations characterised by the concept of 'community without propinquity' (Webber 1964).

According to the idea of the 'compact city', for example, inner cities are the most suitable locations for a sustainable urban development. High density housing development (in particular cheaper, rented social housing), proximity to employment opportunities, a good mixture of shopping and other facilities, lower rates of car ownership and a denser network of public transport, all are in accordance with the British government's sustainability criteria. But the fact remains that inner cities are today the least popular locations for living and working. Concentration does not necessarily imply more interaction, unless these institutional structures are closely related to each other in a set of interconnected institutional webs.

Sustainable development is not something that can be 'summed up' or 'averaged out' for nations, regions, cities, or for individuals. While a macro-analysis of London's institutional structures provides an overview for the planning of sustainability, a microaccount focusing on the intrinsic links between different daily moments is crucial for its successful implementation. This is the rationale behind Chapters 3, 4 and 5. The lesson to be learned is that numerical parity and spatial proximity alone are insufficient to establish the necessary connections between urban institutions.

By virtue of the human scale embedded in the coordination of everyday household life, qualitative matches between job skills, housing features, retail structures and transport systems are the key to integrating institutional structures. Unless the qualitative perspectives and the microcontexts of institutional connections are fully addressed, a nominal approach, or 'spatial fix', is unlikely to achieve the goal of sustainable development. The decline of the inner cities and the unsuccessful attempts to redevelop these areas clearly illustrate the defect of the wholesale, nominal approach of urban reconcentration. In other words, it is insufficient to focus on space or urban forms alone. Such an approach ignores the necessary internal connections between institutional structures. Rather, what is more important is to restore the human scale of institutional connections by addressing the time–space links between everyday life and the overall urban structure.

To achieve this goal, first and foremost, the view that planning is a problem-solving activity needs to be revised. By virtue of the dualities between the routinised daily practices and the institutionalised urban structures, it is more appropriate to see planning as an activity of problem *setting*. It is an interactive process through which the problems to be dealt with, as well as the ways to cope with them, are defined together with the conditions that possibly make the interacting partners engage in a joint action (Crosta 1990; Myerson and Rydin 1994).

In so doing, the concept of urban social sustainability can be considered to be a guiding principle for the planning of a sustainable city. This concept stresses the necessary interconnections between the reproductive development of individual livelihood and the created environment of institutionalised urban structures in a wider context. It links the developmental needs of society as a whole with the very material basis of the natural environment, or the fundamental linkage between people and their environments. Following this line of thought, it would be counterproductive to 'plan' a sustainable city based on an unrealistic assumption of urban containment by focusing only on the existing urban boundaries. Rather, it should be examined in a wider regional context by addressing the necessary connections between institutions.

Economic activities, environmental impacts and social relations have little respect to the borders of the cities, especially those artificial boundaries defined for administrative bodies. In fact, one of the defining characters of modern cities is the close linkage between cities and their hinterlands, including the flows of people and goods/services, as well as the side effects of unwanted waste and pollution across the city borders. Within this broader regional context there would be scope for diverse patterns of development to ensure the flexibility required for coordination of an increasingly fragmented household life. This is especially important for disadvantaged households, whose needs are more likely to be neglected by market mechanisms alone. Breheny's notion of the 'social city region' is useful in this regard. It suggests a starting point for institutional integration in a wider regional scale that reserves the diversity and flexibility of coordination between household dynamics and institutional structures (see Breheny 1993; Breheny and Rookwood 1993). In other words, sustainable urban development requires a combination of different kinds of development, including compact development, low-density development, decentralised concentration, and other types of development, to coordinate the time–space disparities between different daily moments for different households.

But simply to extend the scale of analysis to a regional level is not enough. As previous chapters illustrated, labour markets, housing markets, retail developments, transport systems and other urban institutions are closely related to each other. They are not discrete entities. Appropriate quantities and qualities of housing need to be provided in areas where jobs are created and suitable jobs must be created in areas where people live. Facilities and services must be available at convenient times and places. Most importantly, a comprehensive transport system is required that adequately balances public and private transport to ensure easy access to sites of employment, residence and other facilities and services.

As might be expected, this is not an easy task for policy-makers. Central government or local authorities alone are unable to achieve this goal. Rather, it requires a great deal of integration and coordination within and between institutional boundaries, between administrative hierarchies, as well as between public and private sectors. Moreover, governments need to provide an enabling environment for the coordination of individual household lives, in particular for those disadvantaged groups that are more likely to live against the grain of cities. It can be argued, however, that greater understanding of the dynamic processes behind the coordination of everyday household life provides a useful framework for a more holistic approach to the planning of a sustainable city for both central and local governments.

Conclusions

Throughout this chapter we have argued that the deeper causes of unsustainable trends can be traced to the underlying logic of industrial capitalism – to an endless expansion in both production and consumption via a careless exploitation of environmental resources – and its time–space implications for the reproduction of both individual livelihood and the production system as a whole. At the beginning of the twenty-first century, capitalist expansion and industrial exploitation are reaching an unprecedented level that is characterised by the 'globalisation' of almost every field of production and consumption. Now is the critical point for a reconsideration of its practical implications for both everyday life and the environment as a whole.

Mainstream concepts of physical sustainability that focus exclusively on the trade-off between economic growth and environmental protection lack sensitivity to the internal, social aspect of sustainability. Arguably, this dimension is the ultimate goal of sustainable development. While global environmental crises threaten the survival and prosperity of both current and future generations, sustainability based on the external connections between economic development and the natural environment, such as the neo-classical economic view of 'valuing the environment', are likely to widen, rather than to narrow, the inequalities between nations, regions and social groups by virtue of the capitalisation of both environmental and human resources. Though rightly addressing the relationship between human society and its material basis in the natural environment, this approach ignores what we regard as the very nature of sustainable development: its social embeddedness.

Most importantly, the means of development should not be conflated with the goals of development. An overall sustainability cannot be achieved by an external integration of physical sustainability alone. It is an integrated issue concerning the very interdependence of people's livelihood and their environments. To establish a harmonious interdependence between people and the environment, we need to bring together concerns about the environment, economy and society, linking both global and local interests, reconciling the conflicts between developed and developing countries, and reducing the inequalities between the haves and the have-nots.

Sustainability issues are both complex and dynamic, cutting across the boundaries of natural science and social science. In some sense, the growth of environmental science, characterised by greater interdisciplinary collaboration, has successfully broken the barriers of previous academic divisions of labour that helped shape current trends of unsustainable development. However, sustainability debates in the tradition of environmental studies only touch the external, physical relations of sustainability issues. While arguing that the social origins of current environmental problems provide a deeper explanation of sustainability issues, this chapter concludes that efforts should be made to develop a new framework of study that addresses the internal, social dimension of sustainability. This framework would provide an integrated framework to coordinate the fragments of everyday life and mismatched institutional structures. In so doing, it would be possible to recast current urban problems into a more realistic framework of analysis concerning the situatedness of people's livelihood within the created environment. Only if we establish a strong grasp of sustainable development based on the necessary connections between people and their environments can we manage the economy, society and environment in ways that *meet the needs of the present without compromising the ability of future generations to meet their own needs.*

Note

1. The first report of this series is *Blueprint for a Green Economy*, a book commissioned by the then Department of Environment (DoE) and prepared by the London Environment Economics Centre (LEEC). The latter is a joint venture established by the International Institute for Environment and Development (IIED) and the Department of Economics at University College London (UCL).

Chapter 7

Widening the web

The journey, not the destination

Our everyday life seems to be governed by a binary logic; we are either *here* or *there*. Being 'in-between' is to be not recognised. Set against this we might place a conception of everyday conduct drawn from Zen philosophy: the journey, not the destination, is the thing. Our point in this book can perhaps be best aligned to this last statement. In a sense we are always 'in transit(ion)' from one state to another. If we adopt the binary rationality and only perceive, or register, the arrivals and departures then we are apt to miss out on a significant aspect of everyday experience.

We have sketched out the simple and obvious point that people take responsibility and control of their own actions; however, these are seldom under ideal or desirable conditions. Nevertheless, people do 'go on', they – we – make the best of what we have. People have to regularly come up with extraordinary solutions to solve the most mundane of problems: holding down a job, finding somewhere to live, and having a social life.

The point that we have made several times in this book is that the overwhelming tradition of both the analysis and the governance of cities, the stage for this action, has adopted the binary perspective. While there have undoubtedly been many creative solutions to the problems of housing, work and movement in cities, they have consistently worked with an idealist notion of what people should do (and what the city should be like), rather than addressing what people actually do (and how the city is).

We think that a shift of perspective is required such that the majority of our city lives are not characterised as deviant or irrational, nor are they consigned to 'secrecy'. It is not that we want to simply 'turn the tables'; we argue that urban policy-makers and analysts are consistently missing the target with their policy implementation because they have made themselves blind to many of the issues. The result, we claim, is that the policy recommendations do not impinge upon everyday practice, or, where they do, people are powerless to respond to them. Environmental issues are a very clear case in point.

Many people are keen to reduce pollution in cities, they accept that cars play a significant role in this, and they agree that car use should be diminished. However, they will not give up their car. Why? Traditional economists would have us believe that this is the 'free rider' problem; we hope that somebody else will make the sacrifice. We disagree. Evidence supports the point that people are trapped in a particular combination of transport modes. However, in this book we have gone further than the standard response that an alternative transport mode has to be provided, with appropriate convenience, before people will give up cars. We argue that transport cannot be seen in isolation from housing, work and more general social reproductive activities, and their embeddedness in either the household or wider community dependency networks.

It is only by studying and analysing where people actually are, and which sets of dependencies and relationships they are enmeshed in, that we can begin to understand, let alone consider what might facilitate changes in, their circumstances. We hope that the analysis presented in the main part of this book will go some way to illustrating this point for one city: London.

The aim of this chapter is twofold: first, to draw together the threads of the argument and, second, to draw out some broader implications for thinking about the problems beyond our case study material. We begin by recapping the general argument and its relation to debates about sustainable and compact cities. As a provocation, we include a vignette of the home–work relationships in Portland, Oregon, located in the Pacific north-west of the USA. Portland has been frequently cited as an inspiration for the application of urban design approaches to compact city policy (see Urban Task Force 1999). Finally, we highlight what we feel are the lessons that might be carried over into future research and policy-making.

Recapping the themes of the book

We began this book with an illustration of the everyday innovative solutions that people make to the problem of coordinating daily life. This may have, at first sight, seemed a long way from the elaborate debates about urban sustainability and home–work balance associated, as they are, with transport modellers and urban designers. However, we trust that readers will now appreciate

our starting point. Moreover, the argument will have been taken that traditional solutions to these problems are, at best, 'sticking plasters' on more
fundamental 'illnesses' and, at worst, such stop-gap solutions may actually
exacerbate the problems by encouraging further what John Adams (1999)
evocatively terms 'hypermobility'.

Our arguments are rooted in a particular understanding of urban development that we have elaborated. While we do see the home–work imbalance as
one of spatial dislocation, we recognise this as a superficial reading. Along
with critics of the urban-design-dominated compact city concept, such as
Breheny (1992b, 1999), we stress the broader trends of industrial restructuring and counter-urbanisation that have occurred in most developed nations.
Coupled to this economic restructuring we also note the importance of particular forms of housing provision; especially with respect to owner-occupation.
Allied to this is the issue (in the UK in particular) of the low rate of new house-
building, coupled with increased demand for smaller and single-person dwellings. The result is a basic infrastructural mismatch between residential and
industrial locations. The Urban Task Force's well-meaning solution of new
house-building on brown field urban sites seems somewhat paradoxical here.
Industrial restructuring led to relocation out of cities, and the abandonment
of inner-city sites. To develop these for housing makes sense in economic
terms but is hardly going to provide appropriate housing as it is built exactly
on the site vacated by industry! Of course, such a dislocation can be overcome, at a cost, through transport provision. There is a parallel here with the
story of suburbanisation; though in this case it is the (service and manufacturing) industry that is moving out of cities, or from larger to smaller cities. This
structural need to travel is a cost in both time and energy. We have also noted
that it is a cost that is not born equally by all; the relative impact on the lower
paid is more significant. Not surprisingly, the hypermobile society is exclud-
ing for those who cannot afford to travel. Exclusion may not only take the
form of unemployment, but can also include access to retailing, schooling and
social life.

There is a further paradox with regard to planning new housing at higher
densities for smaller household units. In effect, two competing energy reduc-
tion initiatives are being pursued. On the one hand, initiatives promoting
telecommuting (information and communications technology installed in
homes or local telecommuting centres) are favoured for environmental reasons
(Handy and Mokhtarian 1995). At the same time, initiatives to bring about
greater spatial integration of home–work functions feature widely in planning
documents. Add to this the assumption that rising numbers of single-person
households suit smaller homes at higher densities in central urban locations.
One neglected consequence of these combined initiatives is the trend for
telecommuting to effectively demand larger houses to accommodate personal
home–work space in decentralised locations. Telecommuting also frequently
entails duplication of office equipment and working space, where employees
work one or two days a week at home and for the remainder commute to a
separate office site for purposes of face-to-face supervision or client contact.

Photograph 7.1   1970s high-density social rented housing in London (Source: Helen Jarvis)

Moreover, as we stressed in earlier chapters, household structures and housing choice associated with these are not static. Given these competing pressures, it is unlikely that significant land or energy efficiency gains necessarily flow from the shift towards more prolonged or frequent single-person life-course transitions. The potential conflicts facing decentralised telecommuting and compact city initiatives remain significant in this regard (Gillespie 1992; Graham 1997) (see Photograph 7.1).

Extending on Breheny and his co-authors' work, we also agreed with the sentiments expressed in Allen and Hamnett's (1991) work on the segmented and interrelated nature of housing and labour markets. The co-location of housing and employment is insufficient unless it is the appropriate housing for the person in that work. The complex segmentation of housing and labour markets now makes it tremendously difficult to resolve home and work in a collective fashion. For the most part, we have moved beyond the age of the factory town, or even the industrial suburb, dominated by a single employer, or a variety of similar employers. As we noted, it is now commonplace in major cities for inner cities to be facing a crisis of 'servicing' workers (teachers, cleaners, nurses, etc.). These segments of the labour market are poorly paid and thus have little or no access to the housing markets of urban areas. Consequently, demand is met through long commuting. In our reformulation we have stressed the importance of retail restructuring and the changing nature of school provision that confound the possibility of living a 'compact life' in any city.

The main point that we sought to demonstrate, and to draw out of our analysis, was that multiple market, or institutional, coordination and segmentation was only part of the story. Our point here has been to emphasise that people are not caught in this maze of decision and coordination by themselves. Invariably, people are locked into coordinating problems of other members of their household. This embeddedness spreads outwards to other forms of social support and social reproduction. We would stress that the 'home–work' balance is all about 'production'; it ignores the work (and formidable coordination issues associated with it) of reproduction. Unless these two are seen as two sides of the same coin then coordination problems will abound. This led us to conclude that there was a huge amount of secret or hidden work that goes on when coordinating and reproducing everyday life, of which the home–work relationship is part, but not the dominant part. The growth of multi-worker households as well as the complexities of coordination with retail and educational provision has made for a formidable problem set. It is our view that these households may be seen as a nexus of this problem and, as such, they are perhaps the best places to begin an investigation to understand the nature of the problem, and what, if any, types of solutions may have an impact.

It seems clear from these arguments and the evidence that we have presented in previous chapters that regulating the flow of movement through pricing or infrastructure provision is likely to have a minimal impact. In fact, it may exacerbate the already stressful time–space balance that most people are engaged in. Any solutions must engage with where people are in their everyday movement patterns and dependencies; solutions, if they are to be effective, must work with, and take inspiration from, the multitude of local solutions that individuals and households regularly come up with. Moreover, it is very clear that a major contributory factor to social exclusion and poverty in contemporary societies is (im)mobility. More movement does not seem to be the solution; likewise, banning movement is not a solution either. We must work towards solutions that allow us to live our lives with less movement and fewer coordination challenges. This could be one route towards a more inclusive, and more sustainable, society in both social and environmental terms.

The secret life of Portland, Oregon

Continuing our concluding reflections on what works and what does not, we will now briefly turn our attention to Portland, Oregon. This is a provocative choice as Portland has been held up, and used, as an exemplar of the application of urban design principles to this problem. The popular movement towards 'urban village' planning policies draws heavily on the experience of Portland, which is held by many US states as a paradigm for integrated land use, transport and environmental planning (Abbott 1994; Adler 1994; Kelbaugh

Photograph 7.2   1990s high-density 'urban village' private-sector housing in Portland
(Source: Helen Jarvis)

1997: 27). Applying our analytical perspective, we have a number of caution-ary comments.

At first glance, Portland's story of restructuring – charting a significant shift from primary- and agricultural-sector production to service-industry growth – mirrors that described for advanced economies elsewhere. Yet the plot turns when it is recognised that labour market expansion rides on the back of strong resistance to established patterns of growth. While representing a core provincial city, which has transformed the fortunes of the region, Portland does not submit easily to the costs and benefits generally associated with full participation in the global market. On the one hand, the metropolitan area is currently experiencing significant net in-migration, particularly from California. Young professionals both pursue and effectively attract new start-up companies. On the other hand, the city continues to present an image of parochial progress, the 'big little city', emulating the civic facilities and cultural capital of a contemporary city while retaining the sense of community and 'feel' of a small town. Indeed, a significant number of the new in-migrants select Portland as a location of choice to escape the congestion associated with other west-coast cities. The result, of course, is that Portland is now experiencing unprecedented pressures on land for new housing and greater infrastructure provision (see Photograph 7.2). House prices have escalated rapidly as has the cost of living generally, though still trailing both San Francisco to the south and Seattle to the north.

Portland promotes itself as a compact city, placing great emphasis on environmental quality and a pronounced 'anti-car' image. City planners and policy-makers pursue contained urban growth, the development of 'balanced

communities' and close reconciliation of employment and residential locations. In many respects, Portland can be said to provide a somewhat atypical
metropolitan case in which to observe local contexts of global market restructuring. Here, the structural constraints and uneven development most typically associated with contemporary US cities and advanced world cities such as
London, Sydney and Tokyo are significantly modified by policies directed
towards urban containment. The argument is that this militates against the
most negative attendants of growth: congestion and poor environmental quality, run-down inner areas, spatially scattered and socially polarised housing
and labour market opportunities, urban sprawl and extended commuting patterns. Superficially, this suggests that the spatial coordination of home, work
and community life are more manageable for working parents living in Portland
than the same household type living in cities yielding to more dynamic and
sustained pressures of growth and competition. Indeed, this belief underpins
the 'urban village' and 'urban renaissance' planning goals currently gaining
favour in dynamic, buoyant and more disadvantaged cities alike. Yet, perhaps
surprisingly, household spatial coordination practices in Portland are not so
different to those of large, high-status and particularly buoyant metropolitan
areas. Here too, individuals and households experience significant coordination
difficulties that are not easily resolved by strategic planning alone.

Despite the intentions of planners and policy-makers to concentrate home,
work and transport solutions within a compact urban complex, dual- and
multi-earner households juggling activities in these spheres in the 'liveable
city' of Portland continue to face significant obstacles to coordination over the
long-run. This is because the spatial arrangement and temporal ordering of
everyday activities form one part of a bigger equation. Here, as elsewhere, the
nature, extent and combination of work generated by new flexible employment
practices serve to disrupt individual and group social and spatial mobility.

Analysis of everyday household practices effectively provides one way of
unravelling socially constructed notions of employment flexibility and the
sustainability of household structures and strategies of coordination across
space and time. For instance, issues of coordination extend to everyday strategies that 'plug' chronic 'gaps' in time–space arrangements (such as between
home and child-care locations, working times and school opening hours) as
well as to contingent strategies for managing acute gaps that emerge when a
child is sick or regular child-care or transport provision are unavailable. Existing 'who does what' household research suggests that tensions associated with
the breakdown of daily routines elicit the commonly observed pattern of the
sacrifice or subordination of female spouse employment to male employment.
More particularly, anticipation of this precariousness reinforces persistent gender inequalities and obstacles to the 'strategic' reconciliation of home, work,
transport and child care. Many of the new professional jobs proliferating in
Portland's emerging high-tech industries, for instance, routinely require
employees to travel interstate on business and work to deadlines that demand
unpaid overtime. Other new service-sector jobs are of poor quality, whereby
employment is temporary, working hours short and shifts variable. This means

that solutions have to be found to plug both chronic and acute 'gaps' in time–space coordination, solutions which are frequently embedded in particular places and social relations that function beyond market forces or the reach of strategic planning. Close examination of restructuring in both world city and core regional city contexts reveals the marked potential impact household practices of coordination have on the future structure, growth and cohesion of urban living.

The paradox for Portland (and we argue strongly that this provides an illustration of the failings, not the successes, of the compact city more generally) is that strong public (and private) anti-growth and anti-car sentiments accompany household preference for low-density residential development and continuing reliance on private motorised transport. Unravelling this contradiction requires an understanding of the way urban dwellers are effectively 'torn between identifications' in the coordination of everyday life (Pratt 1998: 26) and in the preferences and decisions they make as household collectives. We can illustrate this by the graphic example offered by the Poulter household.

Steve and Amy Poulter grew up in different parts of California; they met as college students and went to live in Portland shortly after they married. For the Poulters, the ability to 'manage' two full-time jobs and care of their three-year-old child is a particular feature of their centrally located neighbourhood (three miles to the north-east of the downtown area) rather than a general characteristic of Portland's urban containment. Steve explains that they wanted to set up home where they could 'walk and cycle places' and live 'pretty much at the core of things' where 'there are sidewalks and front porches [and] a sense of community'. They claim they benefit from the mutual support of friends in the immediate neighbourhood and that this has gone some way towards compensating for lack of proximity to parents and extended family. Amy points out that strong attachments have formed in the neighbourhood because 'so many [people] in Portland aren't natives'. She is at pains to point out that, like themselves, Californians coming to live in Portland have done so largely to preserve a lifestyle founded on this sense of community:

> [Others] are being asked to be more mobile but [they] are not being compensated, both personally and economically, I mean, with all the technology people are working harder and longer and there's still less [quality of life]. I mean, we just don't know people who are, who talk about their careers in terms of dollars and cents, we just don't know anyone like that, it's not a Portland [thing], you know [. . .] we came here to have a good quality of life.

This narrative suggests a preference for the close reconciliation of home–work–community spatial relations. Yet this couple's everyday practices of coordination tell a different story. While the Poulters extol the virtues of central city living and situate this within an anti-car 'urban village' ethos, both currently require a car to go about their daily work routines. Furthermore, they have changed the location of both employment and child care in ways that increase their need for travel. This is not to suggest 'irrational' behaviour or perverse reaction, but rather demonstration of the competing preferences and identifications that underpin everyday practices, routines and strategies of behaviour.

It has to be asked why the private car continues to dominate circulation in a compact city such as Portland, built on the vision of social and infrastructure integration. One explanation might be that workers are forced to commute longer distances out of, or across, the city because of changes in the location of new employment relative to new and existing residential development. Yet evidence of such a 'mismatch' between jobs and housing appears less marked for 'younger', smaller US urban complexes than was evident for the case of London in Chapter 3.[1] Fewer than one in five workers commute to work outside the county. There is also remarkably little difference in commuting mode and range between central and more suburban residential neighbourhoods. While households in more central neighbourhoods articulate preference for 'convenience', this does not equate to material difference between equivalent households in central and more suburban locations. This suggests that households electing to live in the more central areas of a compact city do not do so as a means of spatially concentrating home and work locations. As with the 'school run' debate in the UK (see Chapter 2), it can be argued that this is evidence that workers' daily lives (that is, their working schedules and patterns and activities of social reproduction) are getting more complex and fragmented. Working parents are packing more trips into busier working days. Moreover, preferences and identities associated with children's education, attachment to place, local social networks and moral cultures typically crosscut those of housing choice, journey to work and personal environmental ethos.

It is not surprising to find that these issues are not confined to Portland. Geraldine Pratt (1998: 39) observes a similar trend in her study of parents employing migrant domestic workers in Vancouver, British Columbia. She notes how a number of middle-class parents in the study area expressed strong commitment to a car-less lifestyle while transporting their children to child-care options outside the local neighbourhood. This apparent contradiction emerges as a function of multiple, competing identifications – residential attachment, personal ethos, local knowledge and naturalised codes of 'good' parenting. Parents in the area exchanged ideas and opinions concerning the quality of local versus more distant child-care options until this exchange 'snowballed' to precipitate significant migration away from neighbourhood solutions. This is further evidence that strong residential attachment and locally embedded networks of knowledge, together with naturalised gender and parenting roles, reproduce multiple, diverse and shifting patterns and identifications of everyday coordination. This tension was first examined in Chapter 3 with respect to the widespread tendency for middle-class parents to transport their offspring to schools outside the local catchment area. On the one hand, these households suggested strong attachment to Inner London housing markets. On the other hand, they articulated the need to select educational qualities that did not match those of residential diversity by indirectly purchasing education from outer urban and suburban markets through the private cost (and negative externalities for society as a whole) of increased transportation.

Some lessons arising from the research ————————————

Ways of seeing

As the title of the book suggests, what we have tried to offer is a new way of looking at the city. Although we have rooted our view in a sophisticated body of social theory, the outcome is quite straightforward: focusing on how people get by in their lives. The social theoretical framework served to mark out the similarities and differences from other work, and to clarify the properties of our chosen lens and to be aware of its blind spots and distortions. The focus on practice is, we believe, a powerful one. It should not be confused with simple empiricism. It is undoubtedly empirical, but informed by social theoretical concepts. These concepts, as we have been careful to point out, try to posit an anti-idealist view of the city. Our point is that such a view, in an unexamined, or un-admitted fashion, lies at the heart of much urban theory and urban policy. Thus, if accepted, our approach creates a new foundation for thinking about these problems. To put it rather bluntly, and to oversimplify, it is the view from the street, not from the 67th floor. This analogy has two dimensions. First, the literal perspective: view from where people are, not the patterns and flows. Second, the implied social and economic distance between those who are consigned to 'the street', and those who work in the well-remunerated jobs located in the ivory towers of the corporate, policy or academic worlds.

Social sustainability

Another theme in this book was inspired by a commonly overlooked sentence in the seminal report of the Brundtland Commission, *Our Common Future* (WCED 1987). The usual 'take home message' from this report concerns sustainability, a notion that is remarkably imprecise and vague; nevertheless, it is one that has found its way into numerous policy statements. However, discussions of environmental sustainability are seldom intimately hinged to social sustainability. In fact, in the Brundtland Report social sustainability is offered as a *prerequisite* for environmental sustainability. Again we can reflect on the way in which a binary opposition is set up in social and environmental issues; added to this is the developmentalist notion that environmental issues are a 'superior order' of human need, to be settled after the basics. Thus, the argument commonly runs, in the developed world social needs can be taken as more or less solved. The point we have to make is that this is far from actual experience. As we have noted time and time again, the complexity, and hard work, that goes into social sustainability makes it an awesome achievement, but one that is undoubtedly fragile.

It is our point that social sustainability is a key problem facing cities in the north, as well as the south. And, as the Brundtland Report notes, it is a prerequisite for environmental sustainability. The example of giving up car use is one aspect of an illustration of this; this is compounded by the extra burden of car usage that we put on the poorer sections of society.

Time and exclusion

Next we turn to the old, and sadly persistent, theme of social exclusion and inequality. We have not approached the book with a conventional perspective on social exclusion, nor have we had this as our central issue, but it does form one of our central concerns. Moreover, we think that our work points to an additional perspective that could be explored by those who make social exclusion their focus. We have sought to explore how the practices of everyday life – making connections and being there – are a considerable achievement that require a lot of work and substantial resources to resolve.

There are a number of clear structural inequalities and mismatches that exclude people from places and opportunities. This much has been a focus of conventional work. Our work reinforces this and demonstrates the multiplication of such problems when they are compounded. What is often required is a substantial resource of social and economic skills, not only in managing budgets, but also engaging in the problems of time-management. Living in cities where mobility and flexibility are the key words requires much of the worker, or aspirant worker. The problems of getting to an interview, or getting to work *on time*, can be as much a barrier as the 'lax timekeeping' of those not used to the labour market, or those excluded from schools. Poor timekeeping is commonly cited as a reason why young persons in deprived areas fail to take up jobs. The simple point is that most unemployment 'blackspots' are relatively close to areas of labour demand; however, getting from home to work is not a simple issue, as we have shown. The same story can be retold for the extra costs borne by the poor by shopping in small corner shops rather than supermarkets. The supermarkets require transport (time and money) to reach them; often it is easier, or only possible, to use the local facilities.

The narrative that we hear expounded about globalisation and global cities by writers such as Sassen and Castells applies *within* the city, as well as between it and others. The islands of exclusion are next to the citadels of success and excess. However, those socially excluded have a smaller and physically delimited life-world, within which there is limited access to resources. 'Being there' is an everyday struggle. Running across, through and beyond these social spaces are those of the socially included (who sometimes may live in spatial proximity to the socially excluded), who have access to resources and are socially and economically connected. This group has few problems in 'being anywhere'. Such a picture undoubtedly oversimplifies the true situation. Our research shows up degrees of exclusion within all households, commonly along cleavages of age and gender (though, of course, there are huge variations that broadly correspond to social class).

Women: being everywhere

If 'being there' is a general problem, it might be said that women as a group have increasingly to 'be everywhere'. The shift in social and economic attitudes of the last half century have given rise to a seismic shift away from the

Photograph 7.3 Changing gender roles and household structures: Microsoft 'permatemp' helping out at home (Source: Steve Ringman)

exclusive male breadwinner household to the multi-earner and sole-woman-earner households. Classically, the social reproductive labour of child rearing and managing the household has been deemed 'women's work'; there is little evidence of this changing, despite the changes outside the home. The term created in the late 1960s is as apt a description as ever: the 'double shift' (one at home, one at work). Likewise, a city's physical form is slow to change. While few neighbourhoods were explicitly designed and planned with the 'woman at home' as norm (see Wagner 1984; Roberts 1991), most bear the marks of wider social and economic norms. Recent innovations include the location of child-care facilities at, or close to, work premises. As yet, school hours have not changed to take account of the new work patterns and social roles.

Earlier chapters identified significant logistical difficulties associated with 'balancing' and 'coordinating' home, work and family life (Witherspoon et al. 1988; Friberg 1993). Households combining two working time schedules and the fixed constraints of child-care opening hours face particular problems, as do those combining two competing careers from a single residential location (see Photograph 7.3). Given that working time and contractual arrangements for individual workers are more varied than can be captured by a single category such as full-time or part-time, temporary or permanent employment (Casey et al. 1997: 19), it logically follows that those combined in two-job households are more complex still. Equally, employment relocation for one spouse can disadvantage the career, pay and prospects of a 'trailing spouse' (most typically female).

Arguably, women's growing participation in paid employment not only impacts upon issues of everyday coordination, but also both reflects and drives new patterns of consumption and social reproduction. This point was made explicit in Chapter 3 with respect to retail development. It was explained how growing female participation in paid labour provided significant impetus to the introduction of late-night opening and Sunday trading because women, who continue to undertake the majority of food shopping, were no longer available to shop in 'normal business hours'. At the same time, extended retail trading along the 24/7 service pattern (24 hours a day, 7 days a week) has generated a disproportionate number of new job opportunities for women, as more women than men are employed in retailing and other service industries involving part-time, anti-social hours and shift work. Equally, it was suggested in Chapter 5 that growth in the number of dual-earner households has played a role in driving the 'treadmill' of consumer expectation, as well as an erosion of male spouse relative earnings and employment security (Jarvis 1997: 527). Elsewhere, we have seen how the material costs and rigidity of the housing market effectively drive a wedge between attempts to match the quantity and quality of jobs and housing in particular metropolitan areas (Johnson 1999).

The grain of the city

We have been keenly aware that the morphology of cities, the arrangement of the buildings and roads, does not offer a 'level playing field' on which the other social relations can be mapped. This should not be read as a 'space matters' plea; of course, there is a uniqueness, a monopoly, of place. If I live here then you cannot. We resist the simplistic notion that social relations can be simply mapped on to spaces; instead, we argue for a co-construction of spaces and social actions. Thus, we note the construction of social areas that combine property rights and economic and social exclusion. Places like Hampstead in north London, or Mayfair in west London. These subtle, and not so subtle, structurings of place (through social and economic practices), as well as the relative connection to transport infrastructures, play a key role in our argument.

Cities grow and develop in a rather haphazard and uneven way. In cities that have a liberal economic governance (for example, London), planning and market structures mesh together over time to produce a particular physical form. These physical structures (roads, factories, shops and houses), their particular situatedness, and their ownership and terms of provision, offer a complex set of limitations on the configuration of work, home and social life. It is thus that we all have to, creatively, make our way through the city each and every day. It is important to note that we are not all in the same position. It is obvious that people have different levels of resources, and different potentials (social, cultural and economic), to overcome coordination problems.

However, our point is that cities do not randomly grow and develop; they do have a pattern, or a structure. We referred to this as the 'grain of the city'. The common pattern is that cities are arranged so that it is easiest for the rich to get around in them; the irony is that they have the greatest resources to overcome difficulties, yet they have least need to draw upon them. Of course, the reverse is also true: getting around, or getting by, is more of a challenge for the excluded; indeed it is a constituent part of being excluded, as the poor must devote a greater proportion of their income on movement.

Beyond this book

It is always tempting to extrapolate from case studies; in effect what we have presented in this book is a rather detailed and partial case study. However, we do have grounds for extending some of the analysis presented here. First, the substantial methodological and theoretical basis developed in this book provides a clear framework for replication of the mode of analysis and the production of comparable findings. Second, despite differences, there are many similarities of process that we might expect to find in other cities. The key differences, things that we could not pronounce upon in detail but which would be the 'place to look' in other cities, can be specified. The list of these will vary in line with the specific local histories of urbanisation, planning regimes, industrialisation, land ownership patterns, housing provision, etc.

One way of making the familiar strange is to carry out cross-national research. We have already noted in Chapter 2 in debates about occupation and employment that social welfare systems can have a crucial modifying impact. If we relate these to the formation and practice of social institutions, such as the household (Esping-Anderson 1990; Duncan and Edwards 1997), we can get an idea how the context is radically shifted. We are, of course, more familiar with the contrasts in housing provision (Ball 1983), employment regulation (Peck 1996) and planning (Hambleton and Taylor 1993).

However, what is characteristic of this work is that it is solely concerned with the institutional policy framework (see Chapter 2). It is generally silent on the actual practices of everyday life that flow from, and between, it. Thus we would stress the need to look closely at the responses to such conditions, and the local solutions created by individuals and households of different social classes. Plus, crucially, what new issues are thrown up in terms of consumption and environmental tensions (who are the new winners and losers?).

Throughout this book a strong case has been made for research that focuses specifically on local contexts of household living, exposing the mundane but essential practices of everyday coordination. Research conducted at this 'practical' level has the potential to amplify and crystallise changing conditions of work, employment and community within and between national and regional boundaries. Cross-national comparative household research offers a particularly powerful route by which to make the connections between patterns and processes of global production and local social reproduction. This is indicated by existing literature focusing on aggregate national and regional data (Morris

1990; Jacobs 1992; Frankel 1997; Duncan and Edwards 1999). Yet little attempt has been made to examine *local practice* in this regard. A contextually rich environment is required for comparative analysis and the opportunity to observe common and diverse elements reproducing the 'secret life' of the world's cities.

One dimension of home–work–family reconciliation not yet afforded due attention, either in socially or spatially directed research, is the mediation of labour market flexibility for a variety of household employment structures; further, the impact that coordination of competing working schedules has on household behaviour in overlapping spheres of restructuring. This is a significant gap in current research and debate, given that the language of flexibility currently extends beyond employment to other spheres of life including housing, transport and child care. Cloaked by ambiguity, notions of flexibility (as with those of sustainability discussed in Chapter 6) are widely promoted as the panacea for all contemporary social problems. The dismantling of 'normal working hours' (for instance, practices such as staggered working schedules, 24-hour services and home-working) is believed to reduce the high social and economic costs of commuting, thus contributing to improved environmental quality (Di Martino and Wirth 1990; Galinsky *et al.* 1993). By this, flexibility might halt further concentration of housing and employment into already densely populated regions and allow smoother space–time reconciliation of separate action spaces. Yet, without examining the inner workings of the household it is impossible to know whether these planning goals represent realistic solutions.

Evidence that there are competing dimensions to the popularly espoused goal of flexible labour needs to be exposed. The way that labour market restructuring is conceptualised, for instance, needs to look beyond the divide between employment and unemployment, to recognise the segmented and fragmented nature of individual and household work portfolios (Jarvis 1997).

Consequently, current emphasis on tackling social exclusion and household income polarisation needs to critically address the ideological promotion as well as the interpretation of flexible labour market practices as a process of labour market restructuring. All forms of employment should work towards providing greater opportunity for shared earning and shared parenting for men and women where this is a preferred goal. At present this is clearly not the case. Working family households increasingly require two incomes to survive. Few would accept the desirability of fathers or mothers spending long hours working away from home, denied contact with their children (Hood 1993). Evidence of high levels of involuntary overtime and long working hours in the UK and the US suggests high levels of stress (and dissatisfaction) for otherwise 'advantaged' high-status, well-paid salaried employees (Schor 1992). Family-friendly policies alone clearly provide insufficient scope for working parents to manage the gaps that open up between home, work and family life in routine or unexpected circumstances. Furthermore, planning and social policy initiatives aimed at simply consolidating home and work appear equally set to fail in this regard. Both further research and policy formation must view household coordination in the round, as the point of origin for the investigation of the social reproduction of urban life.

Note

1. It is interesting to note that the polycentric nature of the Greater Manchester metropolitan area is associated with commuting patterns not dissimilar to US cities. Loftman and Nevin (1995) cite a Manchester City Council report which observes that more than 60 per cent of those working within Manchester city centre travel in each day from another urban centre within the greater metropolitan area.

Appendix

Applying qualitative methods and a household perspective to local contexts in the city

The case studies

In writing this book we have introduced case study material from three separate projects, each employing closely equivalent research methods and design. Here we provide a more detailed discussion of our data collection and analysis to emphasise the unifying principles of a consistently adopted neighbourhood-specific household sampling method. It is instructive to first briefly outline the defining characteristics of each of the three projects.

(1) Four household samples from two contrasting London boroughs

The first project generated 40 in-depth interviews conducted with a cross-section of household composition types (single person, lone mother, nuclear, extended) living in two contrasting London boroughs: Tower Hamlets, located in Inner/east London, and Harrow, located in Outer/west London. Within each borough, two separate neighbourhoods were selected; one reflecting a modest housing stock/income profile, the other reflecting a more prosperous housing stock/income profile. Ten households were interviewed in each of the four neighbourhoods. Households were identified for interview from a doorstep survey of more than 400 households. This provided the basis for selecting a closely equivalent profile of household compositions in each pair of neighbourhoods, as outlined previously in Table 3.1 (page 59).

(2) Working family households interviewed in London and Manchester

The second project generated 30 in-depth interviews conducted with 'nuclear family' households in two comparable residential environments (identified from physical observation as approximately 200 terraced/semi-detached inter-war owner-occupied dwellings with equivalent access to urban/transport amenities); one in Barking, east London, and one in Prestwich, north Manchester. Five interviews were conducted with each of three 'idealised' household employment structure types in each neighbourhood: 'traditional' (single male breadwinner); 'flexible' (one and a half earners); and 'dual' (two full-time earners). A drop-off/mail-back questionnaire provided the basis for selecting an equivalent profile of household structures fitting the subpopulation criteria in each neighbourhood (see Table A.1). The aim of selecting equivalent

Table A.1   Profile of working family households interviewed in two case study areas:

(a) London (Barking)

Interviewee	Male occupation (hours worked) Female occupation (hours worked)	Male/female travel to work mode (journey time)	Child-care provision No. (age) of children
Traditional Type			
Lemon	Bank manager (45) (Bank clerk) Home-maker	train (35) n/a	mother at home 2 (7, 5)
Lampton	Carpenter (37) (Receptionist) Home-maker	car (20) n/a	mother at home 1 (7)
Loader	Lawyer (42) (GP) Home-maker	train (40) n/a	mother at home 3 (4, 2, 1)
Lee	Builder (37) (Bank clerk) Home-maker	car (20) n/a	mother at home 2 (3, 1)
Lever	Manager (45) (Nurse) Home-maker	car (25) n/a	mother at home 2 (8, 6)
Flexible Type			
Lively	Mechanic (44 + OT) Bookkeeper (19)	car (30) bus (15)	unpaid family 4 (9, 7, 7, 4)
Livingstone	Taxicab driver (60) Clerical assistant (28)	car (varies) walk (15)	spouse/unpaid family 3 (12, 10, 7)
Langham	Engineer (40) Secretary (24)	car (20) bus (15)	spouse/unpaid family 3 (12, 10, 4)
Lexington	Local authority manager (36) Local authority officer (19)	car (40) car (20)	paid childminder 2 (5, 3)
Little	Engineer (37) Healthcare officer (19)	car (20) train (15)	unpaid family 1 (2)
Dual-Earner Type			
Lister	Security guard (40) Bank clerk (35)	train (25) car (15)	unpaid family 2 (13, 11)
Leicester	Taxicab controller (50) Finance officer (37)	car (20) public (25)	spouse/dovetail 4 (15, 14, 11, 8)
Dual-Career Type			
Land	Engineer (37) Senior nurse (30)	car (30) car (20)	paid childminder 3 (10, 10, 1)
Lymington	Financial administrator (37) Nursing manager (37)	car (15) car (15)	paid childminder 2 (11, 9)
Linklater	Medical consultant (24) General practitioner (30)	public (30) car (15)	paid childminder 2 (6, 4)

Table A.1   (cont'd)

(b) Manchester (Prestwich)

Interviewee	Male occupation (hours worked) Female occupation (hours worked)	Male/female travel to work mode (journey time)	Child-care provision No. (age) of children
Traditional Type			
Maitland	Graphic designer (40)	car (10)	mother at home
	(Textile designer) Home-maker	n/a	1 (1)
Morley	Engineer (21 × 12, 2 weeks off)	car/plane (varies)	mother at home
	(Childminder) Home-maker	n/a	3 (6, 4, 1)
Mowlem	Engineer	car (varies)	mother at home
	(Civil service officer) Home-maker	n/a	3 (7, 4, 2)
Mannering	College lecturer	car (35)	mother at home
	(Computer analyst) Home-maker	n/a	2 (8, 4)
Masters	Graphic designer	car (20)	mother at home
	(Personnel officer) Home-maker	n/a	2 (2, 3 months)
Flexible Type			
Morris	Lorry driver (42)	car (50)	unpaid family
	Admin. assistant (17.5)	car (15)	2 (8, 4)
Moss	Textile manufacturing worker (40)	bus (15)	spouse/dovetail
	Home help (25)	walk/car (15)	2 (15, 13)
Mellor	Warehouse manager (37 + OT)	car (20)	spouse/dovetail
	Payroll clerk (20)	home/car (10)	1 (4)
Miliken	Security officer (48)	car (20)	spouse/dovetail
	VDU operator	car (110)	1 (3)
Millington	Laboratory supervisor (37)	car (25)	unpaid family
	Healthcare technician (15)	train (20)	2 (11, 9)
Dual-Earner Type			
Myles	Process worker (38 + OT)	car (15)	spouse/dovetail
	Showroom manager (38)	car (115)	1 (11)
Dual-Career Type			
Moore	Engineer (60)	car (15)	paid childminder
	Local authority officer (37)	train (15)	1 (3)
Mallory	Fashion industry promoter (35)	car (varies)	private nursery
	Textile designer (37)	car (15)	2 (4, 3)
Maddison	Insurance sales (38)	train (15)	paid childminder
	Administrative manager (40)	car (20)	1 (2)
Mistry	Photographer/designer (35)	car (varies)	private nursery
	Insurance underwriter (38)	car (20)	2 (8, 5)

subpopulations and residential contexts within contrasting (north–south) metropolitan areas was to examine the contingency of local 'regional cultural' contexts as a factor in explaining the adoption of different 'strategies' by households, of the same type, living in areas supporting contrasting gender regimes.

(3) Working family households interviewed in central and outer Portland

The third project, introduced briefly in Chapter 7, generated 20 in-depth interviews with 'two-wage' working family households living in the US west-coast city of Portland. It represents the first of six metropolitan case studies selected to form part of a larger UK–US cross-national comparative project. Two neighbourhoods – one central (north-east) and one outer (south-west) – were identified from a cluster analysis of Census of Population data. Access to this specific subpopulation of working family households was gained through a carefully targeted postal questionnaire. This was distributed to all households in each neighbourhood, which, from register of voters' data, suggested themselves to be a young family household (two different-sex adults living with no other adults, both born in the period 1957 to 1970). The postal survey was used to refine household selection and to identify those households willing to participate in tape-recorded in-depth interviews.

Neighbourhood-specific household research ────────────

There are several possible ways of collecting household-level primary data. These broadly correspond with workplace/gatekeeper, snowball and residence-based data collection strategies. Each offers discretely different scope for research. First, household groups can be traced from an individual household member identified through an employer or other institutional gatekeeper. This is useful for research focusing on particular occupations, working arrangements or interest groups (such as welfare benefit recipients or trade union members). In this approach, all households interviewed will have at least one member engaged in a common place or practice. Beyond this common element, however, this approach offers little scope to control for household-level parameters such as composition, structure and residential location. Second, households can be identified for interview via a strategy of snowball sampling. In this method, contact is made with a single household matching particular characteristics of interest and this household is then asked to identify others known to it sharing similar characteristics. Thus contact is made through a web of social contacts and informal relations (Lee 1993). This is useful for research focusing on social or kin networks but, again, it results in a geographically dispersed sample about which there is little scope for parametric analysis. Third, households can be identified from their place of residence. We would argue that this is the most appropriate starting point for household-level research focusing on environmental and community variables. A key benefit of a neighbourhood-specific approach is the ability to ensure from geographically-specific secondary data as well as from physical observation that households interviewed share equivalent proximity to transport infrastructure, environmental amenity, 'status' and community profile. This provides a relatively level playing field as background to more detailed observation of intrahousehold discourse and practice.

In the UK, Census of Population data (Small Area Statistics) provides a useful means by which to identify neighbourhoods and household characteristics of particular interest. In case study (1), neighbourhoods were selected from census data to reflect contrasting socio-economic characteristics (occupation/class) for a comparable cross-section of household compositions (single person, couple, family and extended household groups). In case studies (2) and (3), neighbourhoods were selected from census data to ensure broad socio-economic comparability in characteristics such as rate of home ownership, the proportion of households with modern housing amenities, rates of unemployment, ethnic profile and population representation of 'nuclear family' households. The objective was to focus on a relatively homogenous subpopulation of economically active (employed) white 'family' households living in middle-income owner-occupied housing. The selection of a single housing 'type' and vintage also provided a useful platform for comparison, although it is understood that the non-substitutability of housing makes equivalence imperfect.

In each of the case studies a neighbourhood-specific residential survey was conducted as a first step to gain access to the target population. This was used as a means to refine household selection and identify those willing to be interviewed in depth, rather than as a survey in its own right. We considered it important to identify households noting a willingness to be interviewed. This is because rich qualitative data requires households to invest a considerable amount of time and effort in recalling biographical detail. The voluntary nature of participation does not compromise the validity of the data collected since sample collection is confined to predetermined criteria. In these circumstances, Wallman (1984: 47) suggests that 'willingness to cooperate (is) sufficient grounds for selection'. It is quite possible that the methods employed here best suit access to relatively advantaged (educated and employed) households and that an investigation of more disadvantaged population groups would require alternative means of access. For each case study, selection of household was made from those registering their willingness to be interviewed. Interview appointments were subsequently made with willing households for a date and time to suit the availability of adult household members. A letter was sent in each case to confirm the interview appointment and to establish the conditions under which the interview needed to be conducted.

The case studies introduced in this book do not systematically address the role of individual and household class. Instead, emphasis is placed on isolating gender, generation, housing and employment characteristics. By first identifying residential neighbourhoods on the basis of physical observation (housing vintage, type, condition and tenure/status) as well as secondary data for neighbourhood social, economic and demographic profile, a degree of homogeneity is necessarily achieved within each sample of households. Given our emphases, it was not possible to embark on a wider discussion of social and cultural class practices, although such issues emerge implicitly in the intensive biographical research.

Use of in-depth interviewing and 'biographical' data collection

Each of the projects required that housing and employment histories be obtained for all adult household members. A similar semi-structured interview schedule was used throughout, spanning a range of overlapping issues concerning the conduct of everyday life such as: local environment, travel to work, shopping, leisure, social and kin contacts, child-care provision and wider caring responsibilities. In addition, sufficient information was gathered to piece together a 'typical day in the life' of each household, detailing who does what, when, where and with what implications for other households members.

We refer to this combination of semi structured interviews, work histories, chronologies of 'milestone' events and diaries as a 'biographical approach'. Such an approach makes a particularly valuable contribution to housing and labour market research because it is capable of exposing both cross-sectional and longitudinal diversity. By adopting a biographical approach we have been able to contextualise current household structures, employment compositions, social networks, informal economic activities, child-care arrangements and domestic divisions of labour within a more realistic spatial and temporal framework. Short-term fluctuations in divisions of labour are differentiated from longer-term transformations. Martin and Roberts (1984: 2), for instance, note that 'women's lifetime employment is known to be more extensive than cross-sectional data suggests'. Moreover, perceptions of preferred (voluntary) actions are differentiated from those associated with constraint. In effect, a biographical approach has the capability of exploring family households 'situated within everyday life' (Halfacree and Boyle 1993: 338) rather than abstracted as static members of demographic groups and socio-economic types.

Interviewing household members together and apart

In case study (1), all adults in the selected households were interviewed separately. In case studies (2) and (3), working parents were interviewed together. Opinion differs concerning the relative merits and disadvantages of interviewing household members, especially couples, together or apart (Pahl 1989; Valentine 1998). Debate typically hinges on the type of questions asked and the mode of observation. On the one hand, male and female partners may proffer different, perhaps more personal responses to questions relating to household relations when interviewed separately from their partner (Burgoyne 1990). On the other hand, much can be learned from the processes of negotiation that go to make up the coordination of everyday from the joint telling of events (where partners might frequently contradict or reinforce a particular narrative). From a practical point of view it is often difficult to contrive interviews in a home setting with each partner individually. It necessitates separate visits or the banishment of one partner to another room of the house, either of which is likely to reduce levels of cooperation and exposition. Nevertheless, it is apparent from interviews with couples together, where one partner interrupts an interview before it has ended, that narratives shift notably in tenor when spouses are then interviewed apart. In one Portland interview this was particularly revealing. After interviewing Mr and Mrs Paine for an hour, Jon Paine made to leave for a scheduled sporting activity. The interview continued for another half-hour with Liz Paine alone. Liz admitted that her husband had earlier been rather too reticent in describing the way his job, which was more flexible and generally less demanding than hers, requires that he take primary responsibility for looking after their young son and attending to the evening meal during the week. She added that, while he certainly does do all that he says he does, 'he struggles with roles and how much he does [but] he works really hard to try to overcome that' (Liz Paine, 'dual-career', NE Portland). Nevertheless, a biographical approach, by tracing the ordering and interplay of joint and individual housing and employment events, whether conducted with couples together or apart, makes it possible to advance existing theory and debate concerning the effects of economic restructuring on household practices (Morris 1989, 1990; Snaith 1990).

Thematic narrative analysis

The interviews were conducted in such a way as to flesh out key themes (home, residential location, employment, unpaid work, child care, family life, movement) in a flexible, non-linear manner to pursue the train of thought of each respondent (Walker 1985; Holstein 1995). This approach followed well-established principles of grounded theory (Glaser and Strauss 1967; Strauss and Corbin 1998). All subject areas were covered in the course of the interview, together with in-depth probing, regardless of the order in which they were tackled. Clearly, this implies that each interview is a little different from the last, that interviewing is a 'craft' skill (Martin and Roberts 1984) to be honed and perfected along the way and that, to an extent, the latter interviews reflect insights gained from former interviews (Merton *et al.* 1990; Silverman 1993). In setting up and conducting the interviews, then, the intention was to strike a balance between the potentially negative effects of, on the one hand, a rigid adherence to a set interview style, which would prevent this fruitful learning curve and, on the other hand, altering the substance of each subsequent interview so significantly as to negate the possibility of establishing patterns and trends between biographies.

In each case the interviews, each lasting at least one hour, were taped and transcribed verbatim for analysis. Use was made of computer applications suitable for coding, searching and comparing texts (QSR NUD*IST4) where appropriate. It is important to note that while the research did not mobilise socio-linguistic techniques such as discourse analysis (as a formal deconstruction of voices 'behind' the narrative) in the manner currently popular in social psychology (Shotter 1993), an appreciation of this body of literature permeates the conduct and analysis of the interviews. An almost theatrical personal role-play of the interview psychodynamic can be recognised as an important stage of doing qualitative research (Potter and Wetherell 1988). Consequently, we considered it important to record salient observations concerning the conduct of each interview and home setting to ensure that sufficient contextual detail existed for informed transcription, interpretation and analysis of the interview material. In all cases we felt that the interviewees' home provided the 'natural' and appropriate setting, from both an ethnographic and a practical perspective, for household interviews (Robson and Foster 1989). Not only is it the actual site of coordination for home, work and family reproduction, but also it afforded us insight, as participant observers, into domestic practices and arrangements (who answers the door, answers the phone, makes tea, attends to the baby, etc.).

Much of the data sought in the interviews was retrospective in nature and therefore subject to problems of imperfect recall and distortion of past events to suit current opinions. To minimise these potential problems, reference was made to existing research concerning the promotion of respondent recall in life-history analysis. In particular, we drew on lessons learned from the *Working Lives Development Research* by Campanelli and Thomas (1994).

References

Abbercrombie, N. and Urry, J. (1983) *Capital, Labour and the Middle Classes. Controversies in Sociology*, Series 15, London: Allen and Unwin.

Abbott, C. (1994) 'The Oregon planning style', in C. Abbott, D. Howe and S. Adler (eds) *Planning the Oregon Way: A Twenty-Year Evaluation*, Corvallis, Oregon: Oregon State University Press.

Adams, J. (1999) 'The social implications of hypermobility', conference proceedings of *The Economic and Social Implications of Sustainable Transportation, Ottawa Workshop*; working party on pollution prevention, working group on transport, Paris: OECD, (Env/epoc/eppc/t(99)3/Final/Rev1): 99–133.

Adler, S. (1994) 'The Oregon approach to integrating transportation and land use planning', in C. Abbott, D. Howe and S. Adler (eds) *Planning the Oregon Way: A Twenty-Year Evaluation*, Corvallis, Oregon: Oregon State University Press.

Allen, J. and Hamnett, C. (eds) (1991) *Housing and Labour Markets: Building the Connections*, London: Unwin Hyman.

Amin, A. and Thrift, N. (eds) (1994) *Globalization, Institutions and Regional Development in Europe*, Oxford: Oxford University Press.

Anderson, M., Bechhofer, F. and Gershuny, J. (eds) (1994) *The Social and Political Economy of the Household*, Oxford: Oxford University Press.

Anderson, W.P., Kanaroglou, P.S. and Miller, E.J. (1996) 'Urban form, energy and the environment: a review of issues, evidence and policy', *Urban Studies*, 33 (1): 7–35.

Arendt, H. (1970) *On Violence*, Allen Lane: Penguin Press.

Ball, M. (1983) *Housing Policy and Economic Power*, London: Methuen.

Banister, D. (1992) 'Energy use, transport and settlement patterns', in M. Breheny (ed.) *Sustainable Development and Urban Form*, London: Pion.

Banister, D. (1994) 'Transport', in J. Simmie (ed.) *Planning London*, London: UCL Press, 55–70.

Barlow, J. (1990) 'Housing market constraints on labour mobility: some comments on Owen and Green', *GeoForum*, 21: 85–96.

Barlow, J.G. and Savage, M. (1987) *Economic Restructuring and Housing Provision in Berkshire: Some Implications for the Study of Housing and Labour Markets*, Brighton: University of Sussex Urban and Regional Studies.

Barlow, J. and Savage, M. (1991) 'Housing the workers in Mrs Thatcher's high-tech utopia', in J. Allen and C. Hamnett (eds) *Housing and Labour Markets: Building the Connections*, London: Unwin Hyman, 237–51.

Bartelmus, P. (1994) *Environment, Growth and Development: The Concepts and Strategies of Sustainability*, London: Routledge.

Beck, U. (1992) *Risk Society: Towards a New Modernity*, trans. M. Ritter, London: Sage.

Beck, U. and Beck-Gernsheim, E. (1995) *The Normal Chaos of Love*, Cambridge: Polity Press.

Becker, G. (1981) *A Treatise on the Family*, London: Harvard University Press.

Beckerman, W. (1995) *Small is Stupid: Blowing the Whistle on the Greens*, London: Duckworth.

Belle, D. (1982) 'Social ties and social support', in D. Belle (ed.) *Lives in Stress: Women and Depression*, Beverly Hills, Ca.: Sage.

Bertilsson, M. (1984) 'The theory of structuration: prospects and problems', *Acta Sciologica*, 27 (4): 339–53.

Bhaskar, R. (1989) *Reclaiming Reality: A Critical Introduction to Contemporary Philosophy*, London: Verso.

Bianchini, F. *et al.* (1988) *City Centres, City Cultures: The Role of the Arts in the Revitalisation of Towns and Cities*, Manchester: The Centre for Local Economic Strategies.

Bielby, W. and Bielby, D. (1992) 'I will follow him: family ties, gender role beliefs and reluctance to relocate for a better job', *American Journal of Sociology*, 97: 1241–67.

Blaut, J.M. (1975) 'Imperialism: the Marxist theory and its evolution', *Antipode*, 7 (1): 1–19.

Blowers, A. (1993a) 'The time for change', in A. Blowers (ed.) *Planning for a Sustainable Environment: A Report by the Town and Country Planning Association*, London: Earthscan, 1–18.

Blowers, A. (1993b) 'Environmental policy: the quest for sustainable development', *Urban Studies*, 30 (4/5): 775–96.

Bondi, L. (1991) 'Gender divisions and gentrification: a critique', *Transactions of the Institute of British Geographers*, NS16: 190–8.

Bonney, N. and Love, J. (1991) 'Gender migration: geographical mobility and the wife's sacrifice', *Sociological Review*, 39: 335–48.

Bott, E. (1957) *Family and Social Network: Roles, Norms and External Relationships in Ordinary Urban Families* (2nd edn), London: Tavistock.

Bourdieu, P. (1987) *In Other Words: Essays Towards a Reflexive Sociology*, Cambridge: Polity Press.

Bourdieu, P. (1990) *The Logic of Practice*, Stanford, Ca.: Stanford University Press.

Bourdieu, P. and Wacquant, L.J. (1992) *An Invitation to Reflexive Sociology*, Cambridge: Polity Press.

Bowlby, S. (1987) 'Planning town centres for women', *Town and Country Planning*, 56 (9): 261–6.

Breheny, M.J. (ed.) (1992a) *Sustainable Development and Urban Form*, London: Pion Press.

Breheny, M.J. (1992b) 'The contradictions of the compact city: a review', in M. Breheny (ed.) *Sustainable Development and Urban Form*, London: Pion Press.

Breheny, M.J. (1993) 'Planning the sustainable city region', *Town and Country Planning*, April: 71–5.

Breheny, M. (ed.) (1999) *The People: Where Will They Work?* London: Town and Country Planning Association.

Breheny, M.J. and Rookwood, R. (1993) 'Planning the sustainable city region', in A. Blowers (ed.) *Planning for a Sustainable Environment: A Report by the Town and Country Planning Association*, London: Earthscan, 150–89.

Brindle, D. (2000) 'Why London might become like Aspen', *The Guardian*, 30 June.

Bromley, R.D.F. and Thomas, C.J. (1993) 'Retail change and the issues', in R.D.F. Bromley and C.J. Thomas (eds) *Retail Change: Contemporary Issues*, London: UCL Press, 2–14.

Brotchie, J., Batty, M., Blakely, E., Hall, P. and Newton, P. (eds) (1995) *Cities in Competition: Productive and Sustainable Cities for the 21st Century*, Melbourne: Longman Australia.

Brown, M.B. (1974) *The Economics of Imperialism*, London: Penguin.

Brownhill, S. (1990) *Developing London's Docklands*, London: Paul Chapman Publishing.

Bruegel, I. (1996) 'The trailing wife: a declining breed? – Careers, geographical mobility and household conflict in Britain 1970–89', in R. Crompton, D. Gallie and K. Purcell (eds) *Changing Forms of Employment: Organisations, Skills and Gender*, London: Routledge.

Buck, N. (1994) 'Social divisions and labour markets in London: national, urban and global factors', paper from the ESRC London Seminar, LSE, 28 October.

Buck, N., Gordon, I. and Young, K. with Ermish, J. and Mills, L. (1986) *The London Employment Problem*, Oxford: Clarendon Press.

Buckingham-Hatfield, S. and Evans, B. (1996) 'Achieving sustainability through environmental planning', in S. Buckingham-Hatfield and B. Evans (eds) *Environmental Planning and Sustainability*, Chichester: John Wiley & Sons, 1–18.

Burgoyne, C.B. (1990) 'Money in marriage: how patterns of allocation both reflect and conceal power', *Sociological Review*, 38: 634–65.

Burke, T. and Shackleton, J.R. (1996) *Trouble in Store? UK Retailing in the 1990s*, London: The Institute of Economic Affairs.

Campanelli, P. and Thomas, R. (1994) *Working Lives Development Research – Issues Surrounding the Collection of Life-time Work Histories*, London: DoE/Joint Centre for Survey Methods.

Camstra, R. (1996) 'Commuting and gender in a lifestyle perspective', *Urban Studies*, 33 (2): 283–300.

Carling, A.H. (1991) *Social Division*, London: Verso.

Carling, A. (1992) 'Rational choice and household division', in S. Arber and C. Marsh (eds) *Families and Households: Divisions and Change*, London: Macmillan.

Carlstein, T. (1982) *Time Resources, Society and Ecology*, London: George Allen and Unwin.

Casey, B., Metcalf, H. and Millward, N. (1997) *Employers' Use of Flexible Labour*, London: Policy Studies Institute.

Castells, M. (1989) *The Informational City: Information Technology, Economic Restructuring, and the Urban-regional Process*, Oxford: Blackwell.

Castells, M. (1996) *The Rise of the Network Society*, Oxford: Blackwell.

Castells, M. (1997) *The Power of Identity*, Oxford: Blackwell.

de Certeau, M. (1984) *The Practice of Everyday Life*, Berkeley: University of California Press.

Cevero, R. (1989) 'Jobs-housing balance and regional mobility', *Journal of American Planning Association*, 55: 136–50.

Champion, A. and Ford, T. (2000) 'Who moves into, out of and within London? An analysis based on the 1991 Census 2% Sample of Anonymised Records', *Area*, 32 (2): 259–71.

Christie, I. (2000) 'The sustainable family: families, space and time', in H. Wilkinson (ed.) *Family Business*, Demos Collection, Issue 15, London: Demos.

Clark, W.A.V. and Onaka, J.L. (1983) 'Life cycle and housing adjustment as explanations of residential mobility', *Urban Studies*, 20: 47–57.

Cleveland, C. (1987) 'Biophysical economics: historical perspective and current research trends', *Ecological Modelling*, 38: 47–73.

Cohen, T.F. (1993) 'What do fathers provide? Reconsidering the economic and nurturant dimensions of men as parents', in J.C. Hood (ed.) *Men, Work and Family*, London: Sage.

COI (2000) 'Prescott announces £250 million boost for key workers', *Press Release 769*, London: Central Office for Information.

Commission of the European Committees (CEC) (1990) *Green Paper on the Urban Environment* (EUR 12902), Brussels: CEC.

Commission of the European Committees (CEC) (1993) *Towards Sustainability: A European Community Programme of Policy and Action in Relation to the Environment and Sustainable Development*, Brussels: CEC.

Commoner, B. (1973) *Ecology and Social Action,* Berkeley, Ca.: University of California Press.

Conti, S., Malecki, E.J. and Oinas, P. (eds) (1995) *The Industrial Enterprise and Its Environment: Spatial Perspectives*, Aldershot: Avebury.

Cooke, P. (1983) 'Labour market discontinuity and spatial development', *Progress in Human Geography*, 7: 543–65.

Coombes, M., Champion, A. and Munro, M. (1991) 'House price inflation and local labour market influences in Britain', in J. Allen and C. Hamnett (eds) *Housing and Labour Markets: Building the Connections*, London: Unwin Hyman, 157–88.

Crosta, P.L. (1990) 'City and planning, diversity and change', in S. Datta (ed.) *Third World Urbanization: Reappraisals and New Perspective*, Stockholm: Swedish Council for Research in the Humanities and Social Sciences, 267–82.

Crow, G. (1989) 'The use of the concept of "strategy" in recent sociological literature', *Sociology*, 23 (1): 1–24.

CSO (1995) *Social Focus On Women*, London: HMSO.

Dasgupta, M., Frost, M. and Spence, N. (1980) 'Journey to work trends in British cities: 1971–81', *Department of Geography Research Papers 7*, London: LSE.

Davanzo, J. (1981) 'Microeconomic approaches to studying migration decisions', in G. Dejong and R. Gardner (eds) *Migration Decision Making*, Oxford: Pergamon Press, 90–129.

Davis, K., Leijenaar, M. and Oldersman, J. (1991) *The Gender of Power*, London: Sage.

Davis, M. (1990) *City of Quartz*, London: Vintage.

Demos (1995) *The Time Squeeze*, Demos Quarterly Issue 5, London: Demos.

Department of Education and Employment (1997) *Statistics of Education: Public Examinations GCSE and GCE in England 1996*, London: HMSO.

Department of Employment (1983) 'Census of employment final results for September 1981', *Employment Gazette*, 91 (12), Occasional Supplement No. 2, London: Department of Employment.

Department of Employment (1987) '1983 census of employment and revised employment estimates', *Employment Gazette*, 95 (1), London: Department of Employment, 31–53.

Department of Employment (1989) '1987 census of employment', *Employment Gazette*, 97 (10), London: Department of Employment, 540–58.

Department of Employment (1991) '1989 census of employment', *Employment Gazette*, 99 (4), London: Department of Employment, 209–26.

Department of Employment (1993) '1991 census of employment', *Employment Gazette*, 101 (4), London: Department of Employment, 117–26.

Department of the Environment (DoE) (1977) *Inner Area Studies, Liverpool, Birmingham and Lambeth: Summaries of Consultants' Reports*, London: HMSO.

Department of the Environment (DoE) (1990) *This Common Inheritance*, London: HMSO.

Department of the Environment (DoE) (1992d) *PPG 4: Industrial and Commercial Development and Small Firms*, London: HMSO.

Department of the Environment (DoE) (1992e) *PPG 12: Development Plans and Regional Planning Guidance*, London: HMSO.

Department of the Environment (DoE) (1993b) *London: Making the Best Better*, London: Crown Copyright.

Department of the Environment (DoE) (1994a) *Sustainable Development: The UK Strategy*, London: HMSO.

Department of the Environment (DoE) (1994b) *RPG 9: Regional Planning Guidance for the South East*, London: HMSO.

Department of the Environment (DoE) (1996a) *Urban Trends in England: Latest Evidence from the 1991 Census*, London: HMSO.

Department of the Environment (DoE) (1996b) *PPG 6: Town Centres and Retail Developments*, London: HMSO.

Department of the Environment/Department of Transport (DoE/DoT) (1994) *PPG 13: Transport*, London: HMSO.

Department of Transport (DoT) (1994) *Transport Statistics for London 1994*, London: HMSO.

Devault, M.L. (1991) *Feeding the Family: The Social Organisation of Caring as Gendered Work*, London: Chicago University Press.

Dex, S. and Rowthorn, R. (1997) 'Parenting and labour force participation: the case for a ministry of the family', *ESRC Centre for Business Research Working Paper 74*, Cambridge: University of Cambridge.

Dickens, P. (1992) *Society and Nature: Towards a Green Social Theory*, Hemel Hempstead, Herts: Harvester Wheatsheaf.

Dickens, P. (1996) *Reconstructing Nature: Alienation, Emancipation and the Division of Labour*, London: Routledge.

Di Martino, V. and Wirth, L. (1990) 'Telework: a new way of working and living', *International Labour Review*, 129: 529–54.

Douthwaite, R. (1986) *Short Circuit: Strengthening Local Economies for Security in an Unstable World*, Dublin: Lilliput Press.

Duncan, S.S. (1991) 'The geography of gender divisions of labour in Britain', *Transactions of the Institute of British Geographers*, NS16: 420–39.

Duncan, S.S. (1991a) 'Gender divisions of labour', in D. Green and K. Hoggart (eds) *London: A New Metropolitan Geography*, London: Unwin Hyman.

Duncan, S. and Edwards, R. (1997) *Single Mothers in an International Context: Mothers or Workers?* London: UCL Press.

Duncan, S. and Edwards, R. (1999) *Lone Mothers, Paid Work and Gendered Moral Rationalities*, Basingstoke: Macmillan.

Dunning, J.H. (1994) *Globalisation: The Challenge for National Economic Regimes*, Dublin: Economic and Social Research Institute.

Durrschmidt, J. (1997) 'The delinking of locale and milieu: on the situatedness of extended milieux in a global environment', in J. Eade (ed.) *Living the Global City: Globalization as Local Process*, London: Routledge.

Eade, J. (ed.) (1997) *Living the Global City: Globalization as Local Process*, London: Routledge.

Edley, N. and Wetherell, M. (1995) *Men in Perspective: Practice, Power and Identity*, Hemel Hempstead: Simon and Schuster.

Elkin, T., McLaren, D. and Hillman, M. (1991) *Reviving the City: Towards Sustainable Urban Development*, London: Friends of the Earth.

England, K. (1993) 'Changing suburbs, changing women: geographic perspectives on suburban women and suburbanization', *Frontiers: A Journal of Women's Studies*, 14 (1): 24–43.

Ernste, H. and Meier, V. (eds) (1992) *Regional Development and Contemporary Industrial Response: Extending Flexible Specialisation*, London: Belhaven Press.

Esping-Anderson, G. (1990) *The Three Worlds of Welfare Capitalism*, Cambridge: Polity Press.

Etherington, N. (1984) *Theories of Imperialism*, London: Croom Helm.

Fielding, A. (1992) 'Migration and social mobility: south-east England as an escalator region', *Regional Studies*, 26 (1): 1–15.

Finch, J. (1983) *Married to the Job*, London: George and Allen Unwin.

Focas, C. (1985) *Responses to Change in London Transport Fares, Frequency and Quality of Service for Different Socio-economic Groups in Four Areas of London*, London: CILT.

Forrest, R. (1987) 'Spatial mobility, tenure mobility and emerging social divisions in the UK housing market', *Environment and Planning A*, 19: 1611–30.

Fothergill, S. and Gudgin, G. (1982) *Unequal Growth: Urban and Regional Employment Change in the UK*, London: Heinemann.

Frankel, J. (1997) *Families of Employed Mothers: An International Perspective*, New York: Garland Press.

Friberg, T. (1993) 'Everyday life: women's adaptive strategies in time and space' (translated by Madi Gray), Stockholm: Swedish Council for Building Research.

Fukuyama, F. (1992) *The End of History and the Last Man*, New York: Free Press.

Fuller, S. (1988) *Social Epistemology*, Bloomington, Ind.: Indiana University Press.

Galinsky, E., Bond, J.T. and Friedman, D.E. (1993) *The National Study of the Changing Workforce*, New York: Families and Work Institute.

Game, A. (1991) *Undoing the Social: Towards a Deconstructive Sociology*, Milton Keynes: Open University Press.

Garreau, J. (1991) *Edge City: Life on the New Frontier*, New York: Anchor Books.

Geerken, M. and Gove, W.R. (1983) *At Home and at Work: The Family's Allocation of Labour*, Beverly Hills, Ca.: Sage.

Giddens, A. (1981) *A Contemporary Critique of Historical Materialism*, London: Macmillan.

Giddens, A. (1984) *The Constitution of Society*, Cambridge: Polity Press.

Giddens, A. (1985) 'Time, space and regionalisation', in D. Gregory and J. Urry (eds) *Social Relations and Spatial Structures*, London: Macmillan.

Giddens, A. (1990) *The Consequences of Modernity*, Cambridge: Polity Press.

Gillespie, A. (1992) 'Communications technology and the future of the city', in M. Breheny (ed.) *Sustainable Development and Urban Form*, London: Pion.

Girardet, H. (1992) *The Gaia Atlas of Cities: New Directions for Sustainable Urban Living*, London: Gaia.

Giuliano, G. and Small, K. (1993) 'Is the journey to work explained by urban structure?' *Urban Studies*, 30: 1485–1500.

Glaister, S. (ed.) (1991) *Transport Options for London*, London: Greater London Group, LSE.

Glaser, B.G. and Strauss, A.L. (1967) *The Discovery of Grounded Theory: Strategies for Qualitative Research*, New York: Aldine de Gruyter.

Goldsmith, E. *et al.* (1972) *A Blueprint for Survival*, Harmondsworth: Penguin.

Gordon, I. and Molho, I. (1985) 'Women in the labour markets of the London region: a model of dependence and constraint', *Urban Studies*, 22: 367–86.

Gordon, P. and Richardson, H. (1989a) 'Notes from the underground: the failure of urban mass transit', *The Public Interest*, 94: 77–86.

Gordon, P. and Richardson, H. (1989b) 'Gasoline consumption and cities – a reply', *Journal of the American Planning Association*, 55 (3): 342–5.

Gordon, P., Richardson, H. and Jun, M.J. (1991) 'The commuting paradox: evidence from the top twenty', *Journal of the American Planning Association*, 54 (4): 416–20.

Gottdiener, M. (1985) *The Social Production of Urban Space*, Austin, Tex.: University of Texas Press.

Gottlieb, B.H., Kelloway, E.K. and Barham, E.J. (1998) *Flexible Work Arrangements: Managing the Work-Family Boundary*, Chichester: John Wiley.

Gouldner, A. (1960) 'The norm of reciprocity: a preliminary statement', *American Sociological Review*, 25: 161–78.

Government Office for London (GOL) (1995) *London: Facts and Figures*, London: HMSO.

Graham, S. (1997) 'Telecommunications and the future of cities: debunking the myths, *Cities*, 14 (1): 21–9.

Graham, D. and Spence, N.A. (1995) 'Contemporary deindustrialisation and tertiarisation in the London economy', *Urban Studies*, 32 (6): 855–911.

Granovetter, M.S. (1973) 'The strength of weak ties', in S. Leinhardt (ed.) *Social Networks: A Developing Paradigm*, New York: Academic Press, Inc.

Granovetter, M.S. (1985) 'Economic action and social structure: the problem of embeddedness', *American Journal of Sociology*, 91 (3): 481–510.

Granovetter, M.S. and Swedberg, R. (eds) (1992) *The Sociology of Everyday Life*, Oxford: Westview Press.

Greater London Council (GLC) (1975) *Home Sweet Home: A Visual History of LCC/GLC Housing 1888–1974*, London: Greater London Council.

Greater London Council (GLC) (1985) *Women on the Move: GLC Survey on Women and Transport, GLC Women's Committee 2, Survey Results: The Overall Findings*, London: Greater London Council.

Green, A. (1995) 'The geography of dual career households: a research agenda and selected evidence from secondary data sources for Britain', *International Journal of Population Geography*, 1: 29–50.

Green, A. and Owen, D. (1996) 'Severe labour market disadvantage in urban areas', RGS/IBG Annual Conference, University of Strathclyde, 5–8 January 1996.

Green, A.E., Hogarth, T. and Shackleton, R. (1999) *Long Distance Living: Dual Location Households*, London: The Policy Press.

Gregory, D. (1981) 'Human agency and human geography', *Transactions of the Institute of British Geographers*, NS6: 1–18.

Gregory, D. (1994) 'Capitalism', entry in R.J. Johnston, D. Gregory and D.M. Smith (eds) *The Dictionary of Human Geography*, 3rd edn, Oxford: Blackwell.

Grieco, M., Pickup, L. and Whipp, R. (eds) (1989) *Gender, Transport and Employment: The Impact of Travel Constraints*, Aldershot: Avebury/Gower.

Hägerstrand, T. (1975) 'Space, time and human conditions', in A. Karlquvist, L. Lundquvist and F. Snickers (eds) *Dynamic Allocation of Urban Space*, Farnborough: Saxon House.

Hägerstrand, T. (1976) 'Geography and the study of interaction between society and nature', *GeoForum*, 7: 329–34.

Halfacree, K.H. and Boyle, P.J. (1993) 'The challenge facing migration research: the case for a biographical approach, *Progress in Human Geography*, 17 (3): 333–48.

Halifax Building Society (1996) *House Price Indices*, London: Halifax Building Society.

Hall, J.M. (1990) *Metropolis Now: London and Its Regions*, Cambridge: Cambridge University Press.

Hall, P.G. (1989) *London 2001*, London: Unwin Hyman.

Hall, P. and Taylor, R. (1996) 'Political science and three institutionalisms', *Political Studies*, 44 (5): 937–57.

Hambleton, R. and Taylor, M. (1993) 'Transatlantic policy transfer', in R. Hambleton and M. Taylor (eds) *People in Cities: A Transatlantic Policy Exchange*, Bristol: SAUS.

Hamnett, C. (1984) 'The post-war restructuring of the British labour and housing markets: a critical comment on Thorns', *Environment and Planning A*, 16: 147–61.

Hamnett, C. (1996) 'Social polarisation, economic restructuring and welfare state regimes', *Urban Studies*, 33 (8): 1407–30.

Hamnett, C. (1998) 'Social polarisation, economic restructuring and welfare state regimes', in S. Musterd and W. Ostendorf (eds) *Urban Segregation and the Welfare State: Inequality and Exclusion in Western Cities*, London: Routledge.

Hamnett, C. (1999) *Winners and Losers: Home Ownership in Modern Britain*, London: UCL Press.

Hamnett, C. and Randolph, W. (1986) 'The role of labour and housing markets in the production of geographical variations in social stratification', in K. Hoggart and E. Kaufman (eds) *Politics, Geography and Social Stratification*, London: Croom Helm, 213–46.

Hamnett, C. and Cross, D. (1997) 'Social polarisation in London: the income evidence, 1979–1993', paper presented at the ESRC London Seminar, London School of Economics, 30 May 1997.

Handy, S.L. and Mokhtarian, P.L. (1995) 'Planning for telecommuting. Measurement and policy issues', *Journal of the American Planning Association*, 61 (1): 99–111.

Hanson, S. and Pratt, G. (1991) 'Job search and the occupational segregation of women', *Annals of the Association of American Geographers*, NS81 (2): 229–53.

Hanson, S. and Pratt, G. (1995) *Gender, Work and Space*, London: Routledge.

Harbinson, S. (1981) 'Family structure and family strategy in migration decision making', in G. Dejong and R. Gardner (eds) *Migration decision making*, Oxford: Pergamon, 225–51.

Harre, R. (1979) *Social Being: A Theory of Social Psychology*, Cambridge, Mass.: Blackwell.

Harris, N. (1983) *Of Bread and Guns: The World Economy in Crisis*, London: Penguin.

Harrop, A. and Moss, P. (1995) 'Trends in parental employment', *Work, Employment and Society*, 9 (3): 421–44.

Hartman, H. (1981) 'The family as the locus of gender, class, and political struggle', *Signs*, 6: 366–94.

Hartsock, N.C.M. (1983) 'The feminist standpoint: developing the ground for a specifically feminist historical materialism', in S. Harding and M.B. Hintikka (eds) *Discovering Reality*, Dordrecht and London: Reidel.

Harvey, D. (1989a) *The Urban Experience*, Oxford: Blackwell.

Harvey, D.W. (1989b) *The Condition of Postmodernity*, Oxford: Blackwell.

Harvey, D.W. (1996) *Justice, Nature and the Geography of Difference*, Oxford: Blackwell.

Haughton, G. (1990) 'Segmented labour markets', *Area*, 22: 339–45.

Haughton, G. and Hunter, C. (1994) *Sustainable Cities*, London: Jessica Kingsley.

Healey, P. and Shaw, T. (1994) 'Changing meanings of "environment" in the British planning system', *Transactions of the Institution of British Geographers*, NS19: 425–38.

Henwood, M., Rimmer, L. and Wicks, M. (1987) 'Inside the family: changing roles of men and women', Family Policies Study Centre, Occasional Paper No. 6, London: FPSC.

Hetherington, P. (2000) 'Minister pledges £250m home aid for key workers', *The Guardian*, 25 July.

Hewitt, P. (1996) 'The place of part-time employment', in P. Meadows (ed.) *Work Out – or Work In? Contributions to the Debate on the Future of Work*, York: Joseph Rowntree Foundation.

Hiller, D.V. (1984) 'Power, dependence and division of family work', *Sex-Roles*, 10 (1): 1003–19.

Hills, J. (1995) *Joseph Rowntree Foundation Inquiry into Income and Wealth*, York: Joseph Rowntree Foundation.

Hindess, B. (1988) *Choice, Rationality and Social Theory*, London: Unwin Hyman.

Hirsch, B. (1981) 'Social networks and the coping process: creating personal communities', in B. Gottlieb (ed.) *Social Networks and Social Support*, Beverley Hills, Ca.: Sage.

Hochschild, A.R. (1997) *The Time-bind: When Work Becomes Home and Home Becomes Work*, New York: Metropolitan Books.

Hochschild, A.R. with Maching, A. (1990) *The Second Shift: Working Parents and the Revolution at Home*, London: Piatkus.

Hodge, I. and Whitby, M. (1981) *Rural Employment: Trends, Options and Choices*, London: Methuen.

Hodgson, G.M. (1988) *Economics and Institutions: A Manifesto for Modern Institutional Economics*, Cambridge: Polity Press.

Hodgson, G.M. (1993) *Economics and Evolution: Bringing Life Back into Economics*, Cambridge: Polity Press.

Holding, V. and Tate, J. (1996) *Meeting the Demand for Housing Land and the Quest for Sustainability: Are New Settlements Sustainable?* Research paper No. 16, Faculty of the Built Environment, Birmingham: University of Central England.

Holstein, J.A. (1995) *The Active Interview*, Thousand Oaks, Ca.: Sage.

Hood, J.C. (1983) *Becoming a Two-job Family*, New York: Praeger.

Hood J.C. (ed.) (1993) *Men, Work and Family*, Newbury Park, Ca.: Sage.

House of Commons, Environment Committee (1994) *Shopping Centres and Their Future*, Vol. 1, London: HMSO.

Jacobs, B.D. (1992) *Fractured Cities: Capitalism, Community and Empowerment in Britain and America*, London and New York: Routledge.

Jacobs, M. (1991) *The Green Economy: Environment Sustainable Development and the Politics of the Future*, London: Pluto.

Jacobs, M. (1993) *Sense and Sustainability: Land Use Planning and Environmentally Sustainable Development*, London: Council for the Protection of Rural England.

Jarvis, H. (1997) 'Housing, labour markets and household structure: questioning the role of secondary data analysis in sustaining the polarization debate', *Regional Studies*, 31 (5): 521–31.

Jarvis, H. (1999) 'Housing mobility as a function of household structure: towards a deeper explanation of housing-related disadvantage', *Housing Studies*, 14 (4): 491–505.

Jarvis, H. (1999a) 'Identifying the relative mobility prospects of a variety of household employment structures, 1981–1991', *Environment and Planning A*, 31: 1031–46.

Jarvis, H. and Russell, W. (1998) 'The use of price in planning for housing: a review of current practice', University of Cambridge, Department of Land Economy Discussion Paper 96, Cambridge.

Johnson, J. (1999) 'Squeezed out: economic boom bypasses teachers in the Bay Area', *San Francisco Chronicle*, 12 July, Business Pages: A13–A16.

Johnson, J., Salt, J. and Wood, P. (1974) *Housing and the Migration of Labour in England and Wales*, Farnborough: Saxon House.

Johnston, R.J. (1989) *Environmental Problems: Nature, Economy and Society*, London: Belhaven Press.

Jordon, B., Redley, M. and James, S. (eds) (1994) *Putting the Family First: Identities, Decisions and Citizenship*, London: UCL Press.

Katz, C. and Monk, J. (eds) (1993) *Full Circles: Geographies of Women Over the Life Course*, London and New York: Routledge.

Kelbaugh, D. (1997) *Common Place: Toward Neighbourhood and Regional Design*, Seattle and London: University of Washington Press.

Kendig, H.L. (1984) 'Housing careers, life cycle and residential mobility: implications for the housing market', *Urban Studies*, 21: 271–83.

Kirby, D.A. (1993) 'Working conditions and the trading week', in R.D.F. Bromley and C.J. Thomas (eds) *Retail Change: Contemporary Issues*, London: UCL Press, 192–207.

Krais, B. (1993) 'Gender and symbolic violence: female oppression in the light of Pierre Bourdieu's theory of social practice', in E. Li Puma and M. Postone (eds) *Bourdieu: Critical Perspectives*, Oxford: Polity Press.

Krekel, R. (1980) 'Unequal opportunity structure and labour market segmentation', *Sociology*, 14: 525–49.

Labour Party, UK (1991) *London: A Strategy for Transport*, London: Labour Party.

Langton, E. (2000) 'Put an end to capital punishment', *The Guardian*, 13 May, Money Section 4–5.

Lash, S. and Urry, J. (1987) *The End of Organised Capitalism*, Cambridge: Polity Press.

Lawton, R. (1963) 'The journey to work in England: 40 years of change', *TESG*, 54: 61–9.

Lawton, R. (1968) 'The journey to work in England: some trends and prospects', *Regional Studies*, 2: 27–40.

Lawton, R. (1977) 'People and work', in J. House (ed.) *The UK Space*, London: Weidenfeld.

Lee, R.M. (1993) *Doing Research on Sensitive Topics*, London: Sage.

Leyhson, A., Boddy, M. and Thrift, N. (1990) 'Socio-economic restructuring and changing patterns of long-distance commuting', *Working papers on producer services No. 15*, Department of Geography, Portsmouth Polytechnic.

Lin-Yuan, Y. and Kosinski, L.A. (1994) 'The model of place utility revisited', *International Migration Quarterly Review*, 32 (1).

Loftman, P. and Nevin, B. (1995) 'Prestige projects and urban regeneration in the 1980s and 1990s: a review of benefits and limitations', *Planning Practice and Research*, 3 (1): 31–9.

London Planning Advisory Committee (LPAC) (1996) *Supplementary Advice: An Integrated Transport Programme for London?* Romford: LPAC.

London Research Centre (LRC) (1993) *1991–92 Annual Abstract of Greater London Statistics*, London: LRC.

London Research Centre (LRC) (1995) *London Housing Statistics 1994*, London: LRC.

London Research Centre and the Department of Transport (LRC/DoT) (1993) *Travel in London: London Area Transport Survey 1991*, London: HMSO.

Maffesoli, M. (1996) *Ordinary Knowledge: An Introduction to Interpretative Sociology*, Cambridge: Polity Press.

Mandel, E. (1975) *Late Capitalism*, London: Verso.

Manser, M. and Brown, M. (1980) 'Marriage and household decision-making: a bargaining analysis', *International Economic Review*, 21 (1).

Martin, J. and Roberts, C. (1984) *Women and Employment: A Lifetime Perspective*, London: DoE/ONS HMSO.

Martin, R. and Rowthorn, B. (eds) (1986) *The Geography of De-industrialisation*, London: Macmillan.

Massey, D. (1984) *Spatial Divisions of Labour: Social Structures and the Geography of Production*, London: Macmillan.

Massey, D. (1994) *Space, Place and Gender*, Cambridge: Polity Press.

Massey, D. (1995) 'Masculinity, dualisms and high technology', *Transactions of the Institution of British Geographers*, NS20 (4): 487–500.

McCormick, J. (1995) *The Global Environmental Movement*, 2nd edn, Chichester: Wiley.

McDowell, L. (1989) 'Women, gender and the organisation of space', in D. Gregory and R. Walford (eds) *Horizons in Human Geography*, London: Longman.

McDowell, L. (1997) *Capital Culture: Gender at Work in the City*, Oxford: Blackwell.

McLennan, G., Held, D. and Hall, S. (eds) (1984) *State and Society in Contemporary Britain*, Cambridge: Polity.

McRae, S. (1986) *Cross-class Families: A Study of Wives' Occupational Superiority*, Oxford: Clarendon Press.

McRae, S. (1989) 'Flexible working time and family life: a review of changes', Oxford: Policy Studies Institute/Pinter Press.

Meadows, D.H., Meadows, D.L., Randers, J. and Behrens, W.W. (1972) *The Limits to Growth*, New York: Universe Books.

Meadows, P., Cooper, H. and Bartholomew, R. (1988) *The London Labour Market*, London: Department of Employment, HMSO.

Meadows, P. (ed.) (1996) *Work Out – or Work In? Contributions to the Debate on the Future of Work*, York: Joseph Rowntree Foundation.

Mehmet, O. (1995) *Westernizing the Third World: The Eurocentricity of Economic Development Theories*, London: Routledge.

Merton, R.K., Fiske, M. and Kendall, P.C. (1990) *The Focused Interview* (2nd edn), London: Collier Macmillan.

Mincer, J. (1978) 'Family migration decision', *Journal of Political Economy*, 86: 749–73.

Mingione, E. (1991) *Fragmented Societies: A Sociology of Economic Life Beyond the Market Paradigm*, Oxford: Blackwell.

Mitlin, D. and Satterthwaite, D. (1996) 'Sustainable development and cities', in C. Pugh (ed.) *Sustainability, the Environment and Urbanization*, London: Earthscan, 23–61.

Mollenkopf, J. and Castells, M. (eds) (1991) *Dual City: Restructuring*, New York: Russell Sage Foundation.

Moorhead, J. (1999) 'Take a walk on the wild side', *The Guardian*, 30 June.

Morris, L. (1988/9) 'Employment, the household and social networks', in D. Gallie (ed.) *Employment in Britain*, Oxford: Blackwell.

Morris, L. (1990) *The Workings of the Household: A US-UK comparison*, Cambridge: Polity Press.

Morris, L. (1991) 'Locality studies and the household', *Environment and Planning A*, 23: 165–77.

Munt, I. (1987) 'Economic restructuring, culture and gentrification: a case study of Battersea, London', *Environment and Planning A*, 19: 1175–97.

Murie, A. and Forrest, R. (1980) 'Wealth, inheritance and housing policy', *Policy and Politics*, 8: 1–19.

Myerson, G. and Rydin, Y. (1994) ' "Environment" and planning: a tale of the mundane and the sublime', *Environment and Planning D: Society and Space*, 12: 437–52.

Nelson, K. and Smith, J. (1999) *Working Hard and Making Do: Surviving in Small Town America*, Berkeley/London: University of California Press.

Newman, P. and Kenworthy, J.R. (1989) *Cities and Automobile Dependence*, Aldershot, Hants: Gower.

Oakley, A. (1974) *The Sociology of Housework*, Oxford: Martin Robertson.

O'Brien, L. and Harris, F. (1991) *Retailing: Shopping, Society, Space*, London: David Fulton.

O'Brien, M. and Jones, D. (1996) 'Family life in Barking and Dagenham', in T. Butler and M. Rustin (eds) *Rising in the East: The Regeneration of East London*, London: Lawrence & Wishart.

O'Connor, K. (1980) 'The analysis of journey to work patterns in Geography', *Progress in Human Geography*, 4: 1–25.

Office of Population Censuses and Surveys (OPCS) (1994a) *Census: Economic Activity*, Vols 1 and 2, London: HMSO.

Office of Population Censuses and Surveys (OPCS) (1994b) *Census: Key Statistics for Local Authorities*, London: HMSO.

Office of Population Censuses and Surveys (OPCS) (1994c) *Census: Workplace and Transport to Work*, Vols 1–3, London: HMSO.

ONS (1991) *Small Area Statistics of the GB census*, London: HMSO.

Osborne, D. and Gaebler, T. (1992) *Re-inventing Government: How the Entrepreneurial Spirit is Transforming the Public Sector*, Reading, Mass.: William Patrick.

Owens, S. (1992) 'Energy, environmental sustainability and land-use planning', in M. Breheny (ed.) *Sustainable Development and Urban Form*, London: Pion, 79–106.

Pahl, J. (1988) 'Earning, sharing, spending: married couples and their money', in R. Walker and G. Parker (eds) *Money Matters*, London: Sage.

Pahl, J. (1989) *Money and Marriage*, London: Macmillan.

Pahl, R.E. (1969) 'Urban social theory and research', *Environment and Planning A*, 1: 143–53.

Pahl, R.E. (1984) *Divisions of Labour*, Oxford: Blackwell.

Pahl, R.E. (1988) 'Some remarks on informal work, social polarization and the social structure, *International Journal of Urban and Regional Research*, 12: 247–67.

Parkes, D. and Thrift, N. (1980) *Times, Spaces and Places*, Chichester: Wiley.

Parsons, T. and Bales, R.F. (eds) (1956) *Family, Socialization and Interaction Process*, London: Routledge and Kegan Paul.

Pearce, D. (ed.) (1991) *Blueprint 2: Greening the World Economy*, London: Earthscan.

Pearce, D. (ed.) (1993) *Blueprint 3: Measuring Sustainable Development*, London: Earthscan.

Pearce, D. (ed.) (1995) *Blueprint 4: Capturing Global Environmental Value*, London: Earthscan.

Pearce, D., Markandya, A. and Barbier, E. (1989) *Blueprint for a Green Economy*, London: Earthscan.

Peck, J. (1989) 'Reconceptualising the local labour market: space, segmentation and the state', *Progress in Human Geography*, 13: 42–61.

Peck, J. (1996) *Work-place: The Social Regulation of Labour Markets*, New York: Guilford Press.

Pile, S. and Thrift, N. (1995) 'Mapping the subject', in S. Pile and N. Thrift (eds) *Mapping the Subject*, London: Routledge.

Pinch, S. and Storey, S. (1991) 'Social polarisation in a buoyant labour market: the Southampton case – a response to Pahl, Dale and Bamford', *International Journal of Urban & Regional Research*, 15: 453–60.

Pleck, J. (1979) 'Men's family work: three perspectives and some new data', *The Family Coordinator*, 28: 481–7.

Porter, R. (1994) *London: A Social History*, London: Hamish Hamilton.

Potter, J. and Wetherell, M. (1987) *Discourse and Social Psychology: Beyond Attitudes and Behaviour*, London: Sage.

Potter, J. and Wetherell, M. (1988) 'Rhetoric and ideology: discourse analysis and the identification of interpretative repertoires', in C. Antaki (ed.) *Analysing Everyday Explanations*, London: Sage.

Powell, W. and Dimaggio, P. (ed.) (1991) *The New Institutionalism in Organisational Analysis*, Chicago: Chicago University Press.

Pratt, A.C. (1991) 'Discourses of locality', *Environment and Planning A*, 23: 257–66.

Pratt, A.C. (1994a) 'Industry and employment in London', in J. Simmie (ed.) *Planning London*, London: UCL Press, 19–41.

Pratt, A.C. (1994b) *Uneven Re-production: Industry, Space and Society*, Oxford: Pergamon.

Pratt, A.C. (1995) 'Putting critical realism to work: the practical implications for geographical research', *Progress in Human Geography*, 19: 61–74.

Pratt, A.C. (1996a) 'Coordinating employment, transport and housing in cities: an institutional perspective', *Urban Studies*, 33 (8): 1357–75.

Pratt, A.C. (1996b) 'Rurality: loose talk or social struggle?' *Journal of Rural Studies*, 12 (1): 69–78.

Pratt, A.C. and Gill, R. (1991) 'Qualitative labour market survey for London Borough of Haringey', for Urban Cultures Ltd.

Pratt, G. (1998) 'Grids of difference: place and identity formation', in R. Fincher and J.M. Jacobs (eds) *Cities of Difference*, London: Guilford Press.

Pred, A. (1977) 'The choreography of existence some comments on Hägerstrand's time-geography and its effectiveness', *Economic Geography*, 53: 207–21.

Pred, A. (1985) 'The social becomes the spatial becomes the social: enclosures, social change and the becoming of place in the Swedish province of Skåne', in D. Gregory and J. Urry (eds) *Social Relations and Spatial Structures*, London: Macmillan.

Punpuing, S. (1993) 'Correlates of commuting patterns: a case study of Bangkok, Thailand', *Urban Studies*, 30: 527–46.

Randolph, W. (1991) 'Housing markets, labour markets, discontinuity theory', in J. Allen and C. Hamnett (eds) *Housing and Labour Markets: Building the Connections*, London: Unwin Hyman, 16–47.

Raven, H. and Lang, T. (1995) *Off Our Trolleys? Food Retailing and the Hypermarket Economy*, London: Institute for Public Policy Research.

Redclift, M. (1987) *Sustainable Development: Exploring the Contradiction*, London: Routledge.

Reeves, D. (1996) 'Women shopping', in C. Booth, J. Darke and S. Yeandle (eds) *Changing Places: Women's Lives in the City*, London: Paul Chapman Publishing.

Rex, J. and Moore, R. (1967) *Race, Community and Conflict*, Oxford: OUP.

Rickaby, P., Steadman, J. and Barrett, M. (1992) 'Patterns of land use in English towns: implications for energy use and carbon dioxide emissions', in M. Breheny (ed.) *Sustainable Development and Urban Form*, London: Pion.

Roberts, J. (1992) 'Do transport policies meet needs?' in D. Banister and K. Button (eds) *Transport, the Environment and Sustainable Development*, London: Chapman & Hall, 248–56.

Roberts, M. (1991) *Living in a Man-made World: Gender Assumptions in Modern Housing Design*, London: Routledge.

Robertson, J. (1989) *Future Wealth: A New Economics for the 21st Century*, London: Cassell.

Robson, S. and Foster, A. (1989) *Qualitative Research in Action*, London: Edward Arnold.

Rorty, R. (1989) *Contingency, Irony and Solidarity*, Cambridge: Cambridge University Press.

Rose, G. (1993) *Feminism and Geography: The Limits of Geographical Knowledge*, Cambridge: Polity Press.

Rossi, P.H. (1980) *Why Families Move* (2nd edn), London: Sage.

Ryle, A. (1975) *Frames and Cages – The Repertory Grid Approach to Human Understanding*, London: Chatto and Windus.

Salt, J. (1991) 'Labour migration and housing in the UK: an overview', in J. Allen and C. Hamnett (eds) *Housing and Labour Markets: Building the Connections*, London: Unwin Hyman, 94–115.

Saltzman Chafetz, J. (1991) 'The gender division of labour and the reproduction of female disadvantage: toward an integrated theory', in R.L. Blumberg (ed.) *Gender, Family and Economy: The Triple Overlap*, Newbury Park, Ca.: Sage.

Sassen, S. (1991) *The Global City: New York, London, Tokyo*, Princeton, New Jersey: Princeton University Press.

Saunders, P. (1995) *Capitalism: A Social Audit*, Buckingham: Open University Press.

Savage, M. and Fielding, A. (1989) 'Class formation and regional development: the service class in the south-east of England', *GeoForum*, 20 (2): 203–18.

Schor, J.B. (1992) *The Overworked American: The Unexpected Decline of Leisure*, New York: Basic.

Schumacher, E. (1973) *Small is Beautiful: Economics as if People Mattered*, London: Abacus.

Schutz, A. (1967) *The Phenomenology of the Social World*, Evanston, Ill.: Northwestern University Press.

Sharp, C., Brownhill, S., Shad, P. and Merrett, S. (1988) 'Housing and demography in London', *Greater London Housing Study Working Paper 3*, Bartlett School of Planning, UCL, London.

Sherlock, H. (1991) *Cities Are Good for Us*, London: Transport 2000.

Shotter, J. (1993) *Conversational Realities: Constructing Life Through Language*, London: Sage.

Silverman, D. (1993) *Interpreting Qualitative Data: Methods for Analysing Talk, Text and Interaction*, London: Sage.

Simmie, J. (ed.) (1994) *Planning London*, London: UCL Press.

Simmie, J. (1994) 'Planning and London', in J. Simmie (ed.) *Planning London*, London: UCL Press, 1–18.

Simpson, W. (1987) 'Workplace location, residential location and urban commuting', *Urban Studies*, 24: 119–28.

Singh, N. (1989) *Economics and the Crisis of Ecology*, 3rd edn, London: Oxford University Press/Bellew Publishing.

Slater, J.R. (1980) *The Demand for Owner-occupied Housing in Greater London, Part 1, Analysis of the Micro-data*, Discussion Paper Series B, No. 63, Birmingham: Faculty of Commerce and Social Science, University of Birmingham.

Slater, J.R. (1981) *The Demand for Owner-occupied Housing in Greater London, Part 2, Analysis of Aggregated Data*, Discussion Paper Series B, No. 74, Birmingham: Faculty of Commerce and Social Science, University of Birmingham.

Smith, G. (1999) *Area-based Initiatives: The Rationale and Options for Area Targeting*, CASE Paper 25, LSE.

Smith, N. (1990) *Uneven Development: Nature, Capital and the Production of Space*, 2nd edn, Oxford: Blackwell.

Smith, N. (1996) *The New Urban Frontier: Gentrification and the Revanchist City*, London: Routledge.

Smith, N. and O'Keefe, P. (1996) 'Geography, Marx and the concept of nature', in J. Agnew, D.N. Livingstone and A. Rogers (eds) *Human Geography: An Essential Anthology*, Oxford: Blackwell.

Snaith, J. (1990) 'Migration and dual-career households', in J.H. Johnson and J. Salt (eds) *Labour Migration*, London: David Fulton.

Soja, E. (2000) *Postmetropolis: Critical Studies of Cities and Regions*, Oxford: Blackwell.

Somerville, P. (1994) 'Tenure, gender and household structure', *Housing Studies*, 9 (3): 329–49.

Sparks, L. (1992) 'Restructuring retail employment', *International Journal of Retail and Distribution Management*, 20 (3): 12–19.

Strauss, A. and Corbin, J. (1998) *Basics of Qualitative Research: Techniques and Procedures for Developing Grounded Theory*, London: Sage.

Taylor, I., Evans, K. and Fraser, P. (1996) *A Tale of Two Cities: A Study in Manchester and Sheffield*, London: Routledge.

Thomas, R. (1973) 'Home and workplace', in P. Hall *et al.* *The Containment of Urban England, Vol. 2*, London: Political and Economic Planning (PEP)/George Allen and Unwin, 295–328.

Thorns, D. (1982) 'Industrial restructuring and change in the labour and property markets in Britain', *Environment and Planning A*, 14: 745–63.

Thrift, N. (1997) 'The still point: resistance, expressive embodiment and dance', in S. Pile and M. Keith (eds) *Geographies of Resistance*, London: Routledge.

Tivers, J. (1985) *Women Attached: The Daily Lives of Women With Young Children*, London: Croom Helm.

Tivers, J. (1988) 'Women with young children: constraints on activities in the urban environment', in J. Little, L. Peake and P. Richardson (eds) *Women in Cities: Gender and the Urban Environment*, London: Macmillan.

Townsend, A.R. (1986) 'Spatial aspects of the growth of part-time employment in Britain', *Regional Studies*, 20: 313–30.

Turner, R.K., Pearce, D. and Bateman, I. (1994) *Environmental Economics: An Elementary Introduction*, Hemel Hempstead: Harvester Wheatsheaf.

Twain, M. (1876) 'Tom Sawyer', in L. Teacher (ed.) (1976) *The unabridged Mark Twain*, with opening remarks by Kurt Vonnegut Jr, Philadelphia, PA: Running Press.

UNDP (United Nations Development Programme) (1992) *Human Development Report 1992*, Oxford: Oxford University Press.

Urban Task Force (UTF) (1999) *Towards an Urban Renaissance* (Chair R. Rogers), London: DETR.

URBED, Halcrow Fox and Donaldsons (1994) *High Accessibility and Town Centres in London*, Romford: LPAC.

Valentine, G. (1998) 'Doing household research: interviewing couples together and apart', *Area*, 31 (1): 67–74.

Van den Bergh, J. and van den Straaten, J. (1994) *Toward Sustainable Development: Concepts, Methods, and Policy*, Washington DC: Island Press.

Vickerman, R. (1984) 'Urban and regional change, migration and commuting – the dynamics of workplace, residence and transport choice', *Urban Studies*, 21: 25–9.

Wagner, P. (1984) 'Suburban landscapes for nuclear families', *Built Environment*, 10 (1): 35–41.

Walker, R. (1985) *Applied Qualitative Research*, Aldershot: Gower.

Wallman, S. (1984) *Eight London Households*, London: Tavistock.

Walmsley, D.J. and Lewis, G.J. (1993) *People and Environment: Behavioural Approaches in Human Geography*, 2nd edn, London: Longman.

Ward, B.E. (1975) *Human Settlements: Crisis and Opportunity*, Ottawa: Ministry of State for Urban Affairs.

Ward, B.E. (1976) 'The inner and the outer limits', *Canadian Public Administration*, 19 (3): 385–416.

Ward, B.E. and Dubos, R. (1972) *Only One Earth: The Care and Maintenance of a Small Planet*, Harmondsworth: Penguin.

Warnes, A. (1972) 'Estimates of journey to work distances from census statistics', *Regional Studies*, 5: 315–26.

Watson, G. (1992) 'Hours in work in Great Britain and Europe', *Employment Gazette*, 100: 539–57.

Watson, S. (1991) 'The restructuring of work and home: productive and reproductive relations', in J. Allen and C. Hamnett (eds) *Housing and Labour Markets: Building the Connections*, London: Unwin Hyman, 136–53.

Watts, R.J. (1991) *Power in Family Discourse*, New York: Mouton de Gruyter.

WCED (World Commission on Economic Development) (1987) *Our Common Future* (The Brundtland Report), Oxford: Oxford University Press.

Webber, M.M. (1964) *Explorations into Urban Structure*, Philadelphia: University of Pennsylvania Press.

Webber, M. (1982) 'Agglomeration and the regional question', *Antipode*, 14 (2): 1–11.

Westlake, T. and Dagleish, K. (1990) 'Disadvantaged consumers: can planning respond?' *Planning Outlook*, 33 (2): 118–23.

Wheelock, J. (1990a) 'Capital restructuring and the domestic economy: family self-respect and the irrelevance of "rational economic man"', *Capital and Class*, 41: 103–41.

Wheelock, J. (1990b) *Husbands at Home: The Domestic Economy in a Post-industrial Society*, London/New York: Routledge.

White, R. and Whitney, J. (1992) 'Cities and the environment: an overview', in R. Stren, R. White and J. Whitney (eds) *Sustainable Cities: Urbanisation and the Environment in International Perspective*, Oxford: Westview Press.

Wilkinson, H. (ed.) (2000) *Family Business*, Demos Collection, Issue 15, London: Demos.

Williams, P. (1982) 'Restructuring urban managerialism: towards a political economy of urban allocation', *Environment and Planning A*, 12: 95–102.

Willmott, P. and Young, M. (1957) *Family and kinship in East London*, London: Routledge and Kegan Paul.

Willmott, P. and Young, M. (1973) *The Symmetrical Family*, London: Routledge and Kegan Paul.

Wilson, W.J. (1987) *The Truly Disadvantaged: The Inner City, the Underclass and Public Policy*, Chicago: University of Chicago Press.

Wilson, W.J. (1996) *When Work Disappears: The World of the New Urban Poor*, New York: Vintage.

Witherspoon, S., Roger, J. and Brook, L. (eds) (1988) *British Social Attitudes: The 5th Report*, Aldershot: Gower.

Wohl, A.S. (1977) *The Eternal Slum: Housing and Social Policy in Victorian London*, London: Edward Arnold.

Wolpert, J. (1964) 'The decision process in spatial context', *Annals of the Association of American Geographers*, 54: 537–58.

World Bank (1992) *World Development Report 1992: Development and the Environment*, Oxford: Oxford University Press.

Worpole, K. (1992) *Towns for People: Transforming Urban Life*, Buckingham: Open University Press.

Wrigley, N. (1993) 'Retail concentration and the internationalization of British grocery retailing', in D.F. Rosemary and C.J. Thomas (eds) *Retail Change: Contemporary Issues*, London: UCL Press, 41–68.

Yeandle, S. (1984) *Women's Working Lives: Patterns and Strategies*, London: Tavistock.

Zucker, L. (ed.) (1988) *Institutional Patterns and Organisations: Culture and Environment*, Cambridge, Mass.: Ballinger.

Zukin, S. (1987) 'Gentrification: culture and capital in the urban core', *Annual Review of Sociology*, 13: 129–47.

Index